My War . . . VIETNAM

My quest for a war . . .
how I arrived at that endeavor,
and some amusing and human interest stories
along the way

Lt/Col Lloyd J. Draayer
USMC, Retired

Permissions

Photography was taken from the author's personal collection and Squadron Cruise Books. Maps and additional photos were through the courtesy of the History and Museums Division, Headquarters, U.S. Marine Corps.

Cover illustrations by Mary Renae Coburn

First printing: January 2012

ACKNOWLEDGEMENT

Lt. Colonel Lloyd J. Draayer served in the Marine Air Corps in the United States Military from 1953 to 1973 and retired after 20 years of service. He was a decorated war hero, serving three tours in Vietnam and flying over 500 combat missions. Lt. Colonel Draayer died at age 70, November 27, 2000.

This is a true story by Lt. Colonel Draayer written in 1989 and never published. He wrote the book for the pilots or "backseaters" who ever flew in the O-1 Birddog, for those who served in the outstanding Squadron VMCJ-1, and for all who lost their lives or served defending our great country. It depicts real events throughout the Vietnam War with a touch of humor about the "other side" of combat.

Lt. Colonel Draayer was my brother, and I am publishing this book in his memory. I edited this manuscript for publication, but I have not changed the content of the story as he wrote it. Some photos that were not in his original writing were added after being found in his memoirs.

Mary Renae Coburn
January 2012

Dedication

To any pilot or backseater
who ever flew in
The O-1 Bird Dog
(and loved her like I did)

Major Lloyd J. Draayer

For those who fight for it,
life has a special flavor
the protected never know.

Author and O-1 Bird Dog

Preface

Since my last tour in Vietnam, I have fathomed numerous books about the conflict. The first ones, written mostly by politicians and dissidents, revolved around America's involvement in the conflict. The second wave, many by military men, detailed the gruesome burden of our ground troops in the South. Then, best of all, came the history books.

Now war is hell, but there is also the zany side of the story. Recalling many humorous incidents I either heard or personally observed, I figured it was time someone published a book articulating the lighter side of war. While researching the focus of the book, I decided there were so many human interest stories that should be told, I included many.

I was a patriotic little cuss, had always figured I was put on this earth to fight a war, and seriously wanted into one, so I have included a few stories about how I finally got to "MY WAR" and a few sidelights on the journey.

The results are a conglomeration of stories gleaned from a poorly kept diary and a fast-failing mind. I've told it like I remembered it.

A word of warning: read this book with tongue in cheek. That is the way it was written.

Contents

Chapter 1

World War II

MY WAR. I'd finally found one . . . Vietnam! I had been searching for a war all my life, and finally, in March of 1968, I was headed into a real shoot-em-up. It proved to be one of the most gratifying experiences in my life.

Lacking a little age, I missed (but was a child of) WWII. Growing up a good Mormon boy in the little town of Clinton, Utah, I was fascinated by the "big one." The one real treat we younguns had growing up on the farm was the weekly Saturday trip to Ogden to browse the toy stores (Newberry, Woolworth, J.C. Penney)—any store that had toy counters. I was just tall enough to see over the counters to view the bright and colorful toys . . . yo-yos, tin soldiers, cars, and airplanes.

Oh, how I wanted to pick up a couple of tin soldiers, point them at each other and go . . . rat-a-tat-tat. I received some of the soldiers a little later in the war, but soon found out there were no replicas of Hitler, Tojo, or Mussolini, and if you couldn't kill the monstrous villains, it wasn't very exciting.

Two-inch-thick "Big-Little" books and the comic books were great reading material, many having patriotic themes. *Superman*, *The Flash*, *The Green Hornet*, *Batman*, and *The Torch* were fantastic. (I also developed a great appreciation for the well-drawn figures of *Wonder Woman* and *Sheena, Queen of the Jungle* . . . my early introduction to shapely women and boobs.)

Family tradition found someone from the neighborhood traveling to Ogden to shop, see a movie, eat dinner, buy an ice-cream cone at Brown's Ice Cream Store, then head for home. The climax for me was the movie theaters. I usually went to town with Grandma and Grandpa Draayer, Mom and Dad, or relatives. They always let me stalk the toy stores for a couple of hours, then sat me down in a theater until it was time to go home. One afternoon my Uncle Claude took me to town, went with me to the movie, and couldn't get me out until the last movie was over and the theater lights came on. Bless his heart—we saw the movie three times. We came home to anxious Mother about midnight.

I loved the movies and still do. My boyhood heroes were John Wayne, Gene Autry, Tarzan, Roy Rogers, and Hopalong Cassidy. Yes, I even remember Buck Jones and Tex Ritter. Still, the movies about the war were my favorites. Walt Disney put out a documentary on the war in Europe that completely engulfed me . . . the maps, assault routes, tactics, etc. Shows the likes of *Bataan* (Robert Taylor), *Thirty Seconds Over Tokyo* (Spencer Tracy) and all the John Wayne war movies just made the hair stand up on the back of my neck. *Sergeant York,* with Gary Cooper, was the greatest movie I had ever seen.

Of course, I loved the *Movietone News*, with its up-to-date portrayal of war in Europe and the Pacific. Even the cartoons had patriotic themes. I lived to find a copy of *Life* magazine with its wonderful pictures of "the front." The radio news right after *Jack Armstrong, the All American Boy*, *The Lone Ranger*, *The Green Hornet*, *The Shadow*, then again following *Amos and Andy*, really got my attention. When Red Skelton hit the airways, he quickly became my favorite. The morning after Mr. Skelton's first broadcast, all the kids in the school bus could say was, "If I dood it, I get a whipin' . . . I dood it!" (Later, while clowning the U.S. Marine Corps Rodeo at El Toro, California, I traveled to Hollywood to appear on television and visited many movie studios to publicize the rodeo. I met Mr. Skelton at one of the studios . . . a real thrill.) My patriotic zeal always flourished when I heard Kate Smith's rendition of . . . "God Bless America."

My interest in flying started by being involved with a civilian light plane crash in our neighbor's field. I was six or seven years old when the pilot had engine trouble and tried to land, first in our field and then in a neighbor's. He didn't notice a small ditch running through Mr. Bybee's freshly plowed field, hit it, and rolled up on his nose. The pilot stayed with us the next couple days, regaling my imagination with tales of "daring-do." I watched in amazement as the wings were taken off the plane. A team of horses hooked up to the plane's tail and pulled it over onto its landing gear. I was thrilled when the whole thing was ramped up into a truck and packed off to the airport.

Later, a friend of my Dad's, Bill Beus, bought a little Stinson airplane and occasionally took us flying. What a thrill! One day when Bill was living in Burley, Idaho, he decided to fly Dad and another gentleman back home to Ogden. The little Stinson only had two seats but had a small luggage compartment behind the seats where he sequestered my Dad and took off—way over the weight of the aircraft's lifting capacity. Near Strevel, Idaho, he couldn't get enough altitude to fly over one of the rolling hills, so he simply landed on the main highway, taxied over the hill, and took off on the other side. (Traffic was rather light in the 30s.)

My eventual entry into the Marine Corps was inspired by my best friend, Burk Clegg. His brother was a Marine, fighting island to island in the Pacific, and was our hero. We knew the "Marine Corps Hymn" by heart (first verse) and sang it at the top of our lungs.

My Grandma Dalton had an orchard in Roy, located on the main highway between Ogden and Salt Lake City. During the war, she used the openings between the trees to rent out trailer spaces to civilian and military people working at the military facilities around Ogden. One renter was a Lieutenant Colonel in the Air Force stationed at Hill Air Force Base. I spent hours in their trailer, pouring through his library and the military magazines he brought home from work. I completely absorbed the stories of the Flying Tigers, the P-40, the pictures and data depicting the P-39 Airacobra, the P-47 Thunderbolt, the B-17, B-25, and the B-24. The aircraft that absolutely enthralled me, however, was the P-38 Lightning. Man, I couldn't wait to grow up and fly one! (There was a joke about the P-38 going around rodeos at the time—seems there is a new beer being brewed called Airplane Beer . . . "drink one and pee 38!") My favorite song was "Coming in on a Wing and a Prayer."

I was educated in the Davis and Weber County school systems, graduating from Weber County High School in May, 1948. During that summer, I was selected to play in the annual Utah High School All-Star Football game held in Salt Lake City at the University of Utah stadium. During our slack time from two-a-day practices, there was a lot of talk about a new draft proposal for young men. The gist of the thing revolved around an opportunity to enter the service for one year and fulfill the usual two-year military obligation. A lot of guys thought this was a great program and were going to sign up after the All-Star game. Not me. When I joined up, it was going to be as an aviator . . . lots of excitement . . . come home every night to a clean bed, showers, good food . . . no muddy trenches or foot sloggin' for me. (I had seen the Van Johnson movie *Battleground*—the Battered Bastards of Bastogne—and wanted no part of trench warfare.)

I had been a pretty good jock during my earlier days and, although receiving tempting offers to attend other colleges, accepted an athletic scholarship to Weber Junior College in Ogden. During my second year at Weber JC, Korea broke out. Some of my good friends, completing their one year in the service, were soon called back in and became "The Battered Bastards of Korea."

I decided to stay in college and accepted a scholarship to the College of Idaho at Caldwell. I figured by the time the Draft Board got around to me, I'd at least have one more year of "educate" in me. While at C of I, a fellow student, a WWII veteran B-25 bomber pilot, enhanced my zeal for fighting "My War" in the air and talked me into taking the USAF Pilot Training Test. Before the test was scheduled, I received my draft notice for a physical and deferment classification. A group of us were loaded on a bus, taken to Boise, deferred or failed physically, and returned. I played football, basketball, and baseball that year, but during the summer, I decided football wasn't much fun anymore. My scholarship hinged on my playing football, so I decided to forgo sports and attend the University of Utah to complete my education. Just before the Fall quarter started, I took the USAF Pilot's Exam, passed, and later received word that all billets were filled.

After finishing my year at U of U, I planned to attend summer school to pick up the few additional credits needed for graduation. Before I registered, however, I ran into an old friend, Jack Oakie, the fine rodeo announcer from Ogden. He talked me into traveling the rodeo circuit with him. I helped feed and water the livestock, worked stock in back of the chutes, and helped load and put flanks on the stock in the bucking chutes.

It was a great, great summer!

Chapter 2

Korea

The Fall of 1952 found me enrolled at the University of Utah to pick up the additional credits needed for graduation . . . still without a commitment from the Air Force. Then, out of the blue, while walking down Main Street in Salt Lake City, I spotted my Air Force vet from C of I. He had a Navy carrier pilot in tow and was headed for a restaurant. I joined them and soon found myself in a hot recruiting war between the two . . . with me as the fish. With interest piqued, I applied for the Navy Pilot tests, was given a free train ride (through the Rockies—not around them) to Buckley Naval Air Station near Denver, took and passed the tests, then came home thinking I'd stay in school right now but take flight training from the first service accepting me or the draft after I graduated.

At the end of the Fall quarter, I took one test on Friday, and figuring I had all weekend to study for the remaining four tests scheduled on Monday, I decided to take in a movie at the Centre Theater downtown. As I entered the theater, much to my surprise, Bill, my Dad's friend, was standing in the foyer. He was living in Nyssa, Oregon, now and was getting ready to grab a train bound for Detroit to pick up a new car at the GM plant. He stated he sure hated to make the trip and slyly suggested, if I went in his place, he would pay all my expenses. Sensing another chance at a new adventure, I took him up on the offer.

On Saturday, while attending a University of Utah football game, I ran into a good friend, newly elected Congressman Doug Stringfellow. How I got to Detroit and back to Utah is a story in itself, but Saturday night found me driving Doug's little Crosley automobile to Washington, D.C., catching a train to Detroit, then driving home through one of the worst winter storms in the history of the West.

I arrived home only to be told Bill was at his home in Nyssa and wanted me to deliver his brand spanking new Buick to Oregon. Why not? I delivered the car, was reimbursed for all expenses, hitchhiked home, and, with school dismissed for the holidays, started planning to go to Southern California with my folks for Christmas. I hadn't given my finals hardly a thought since that fateful Friday night.

A surprise awaited my arrival back in Ogden: a letter requiring my presence at Buckley NAS within five days, where orders to the Naval Cadet Program at Pensacola, Florida, awaited. I was in a quandary, but decided to play it dumb, go to California, open the letter on my return home, then call and explain I had been out of town and unable to pick up my mail.

Upon my return to Utah a week later, I called asking forgiveness and was told to go to the Salt Lake airport the next morning and board a TBM (WWII Navy Torpedo plane) that was flying to Buckley. I dutifully reported, was sequestered in the belly compartment of that old TBM, and over the mountains between Salt Lake and Denver experienced my first flame-out (engine failure). The pilot forgot to switch gas tanks! He restarted the bird, profusely apologized, and we were on our way again. While undergoing my flight physical, my blood pressure was rather high. Luckily, the pilot of the TBM was in the doctor's office. He mentioned our switch-tanks episode. The doc let me lie down for about thirty minutes or so until the meter read okay. I was sworn into the Navy and by six o'clock was on a train headed for Pensacola, Florida. Talk about first class: I had a sleeping compartment . . . all to myself!

To this day, I don't remember the route we took, but I recall switching from a train to a bus somewhere in Georgia. Here I experienced a first impression of my fellow NavCads (Naval Cadets). There were several on the bus, all of them drunk, profane, and making asses out of themselves. Now, to a good ol' non-drinking Mormon boy, this was offensive; however, I did notice that the general public didn't seem to mind . . . after all, it was New Year's Day, 1953.

I promptly settled into Class 2-53, enjoyed my roommates, the sports programs, and, like everyone else, longed for my first airplane. I wasn't a carouser, so when liberty finally became available, I simply went to the movies. One weekend I hit seventeen movies. Saw the Base movie, another one downtown, and the midnight show on Friday night. Saturday I saw two different double features, two single features at other nearby theaters, ending up at another midnight show. Luckily all theater programs changed on Sunday, and I saw the two double features again, the two singles, then the late Base movie. What a waste of time with all those beautiful Florida gals runnin' around.

Ah, Pensacola. Even considering the hectic pace the Navy put NavCads through, we had time for a laugh or two:

We prospective "naval aviators" were given a big dose of PT (physical training). We thought we were in pretty good shape, but one hot, muggy afternoon, we were told to run around one of those immense hangars . . . ten times!

I started off pretty good, passed all of my classmates, and soon found myself way out in the lead. Needless to say, I was flabbergasted when I came off my last lap to find about half the class resting at the finish line. From judicial eavesdropping, I determined they all thought the run was really tough and "that tenth lap was a real killer." Heck, I knew they were all lying. I had led all the way and only done nine laps myself!

One morning right after a strenuous soccer game, for some unknown reason our class was called to the medical facility. We "hurried up to wait" and found long lines of cadets standing in front of several doors. It didn't take long for the word to pass down our line that we were lined up for "shots." The cadet in front of me was a big, strappin' guy who had played tackle for the University of Minnesota. We started talking about the coming event. He stated that shots never bothered him; he had taken his share and wasn't worried in the least.

As we progressed toward our door, he admitted there were a couple of shots that made him a little ill in the past, but he wasn't worried. Not much. As we approached the door, he mentioned casually, "I've known guys to pass out when they take shots."

Upon reaching the door, he made some comment about "needles being short and sharp . . . long and dull" . . . and promptly passed out.

One situation while undergoing training kept me in a continual sweat: saluting. The concept was totally foreign to me. I had great difficulty distinguishing officers from enlisted men while walking around the base. A couple of roommates had mistakenly saluted some enlisted men. That got a great laugh out of me. One day while walking to the gym, I spied a guy in a pickup truck approaching. All I saw was a white cap and a black uniform, so I gave him my smartest salute. It was the damn Base Fire Chief! He slammed on the brakes, pulled over, and laughingly proceeded to give me a lecture on the difference between the Navy and the Fire Department.

After settling into the training command routine at NAS Pensacola, I started thinking about my education again. The head coach during my tenure at C of I, Clem Parberry, had been recalled into the Navy and was a Lt. Commander stationed at Saufley Field, in charge of their athletic program. I renewed our acquaintance. He told me to write one of my teachers at U of U and see if I could arrange to have my final tests sent to him. He would then act as proctor during their administration. I had taken a health class from a Doctor Nemar and felt she might help. She was a tiny little ol' thing, had a wonderful personality, but was tough as nails. Her answer to my letter chewed me out royally, but she stated she'd do what she could to help. Within two weeks, coach called saying all my tests were in.

Bless her heart, Doctor Nemar met with my instructors, talked them into giving her the tests, and sent them to Clem. I took the last test sitting in the back of coach's station wagon on the way to see a football game in Mobile, Alabama. Bama vs. LSU! I was later informed that I had passed the tests and would graduate the following

June. There are some wonderful folks in this ol' world. I made the baseball team while at Whiting Field. During a game against my former coach, I hit a home run to win the game, visually upsetting him. After the game, he came over to me and said, "Draayer, I told my pitcher to throw you nothing but curve balls. You've never hit a curve ball in your life, and look at what you just did to me!"

I can still remember a hop I had with another Cadet, Charlie Burris, out at Saufley Field while undergoing formation flight instruction (two airplanes flying in the same airspace, or in close proximity to each other—the wingman position). We had a thorough brief with our instructor. Fly out to a bridge–highway intersection we frequently used for a rendezvous point. First guy over the designated intersection would lead. After rendezvous, whoever had the lead would go through a series of maneuvers. Once completed, the other student would assume the lead and fly identical patterns. All maneuvers were to be executed following hand signals from plane to plane.

After takeoff, I hit the rendezvous point first, started a slow, left-hand turn, allowing Charlie and the instructor to join on my wing. Charlie joined, the instructor joined and gave me a wave, so I headed out on course. In those days, if your instructor didn't start screaming at you over the radio, you were probably doing a pretty good job. My radio had been silent, so I figured we were getting along handsomely. Charlie and I made probably the best "switch-leads" ever performed in the Training Command, and after completing his portion of our assignments, simply returned to base, landed, and headed for our briefing cubicle. Our instructor was waiting . . . livid!

Apparently, some other instructor had joined up on our flight, realized his mistake, gave us a wave goodbye, and flew off . Our instructor had never joined up nor could he even find us during the two-hour flight. We sure flew a hell of a good hop . . . without him. Charlie and I gave our hop a 4.0 . . . the instructor gave us both a down! We would have to re-fly the hop again, this time with the instructor. (This was the only "down" I received in the Training Command.)

Flight training was fantastic once we got our hands on an airplane. Our class had a few dropouts, but one really sticks out in my mind. I was rooming with this guy at Barrin Field where we took final Carrier Landing Training. The day before qual-day (Qualification), they boated a bunch of us out to the carrier. The USS *Monterey*, a straight deck, was being used at that time. Eight or ten planes would fly out from Barrin the next morning.

Procedure: The fly-out people would shoot two touch-and-gos, then four arrested landings (traps) on the carrier. As each completed his traps, two plane captains would jump up on the wings, one on each side of the aircraft, and, with the engine still running, pull the pilot out one side of the cockpit, stuff the next pilot into the other side, and send him down the deck on take-off. "Switch-pilots" was a hurry-up/shut-up operation.

When I jumped up on the wing to become the next guy, I discovered my roomie in the cockpit, eyes locked in a thousand-yard stare. The plane captains had a little trouble getting him out of the cockpit, but I was soon on my way. After I qualified, I found him in the Ready Room. He looked a little shook to me, so I struck up a conversation. He said, "Lloyd, I'm DOR-ing (Dropped On Request). I'm scared to death. I cannot remember one thing until I saw you waiting to get in the cockpit! I don't remember any landings, take-offs . . . anything." (I couldn't fathom something like that happening, but it sure says something about the Navy training: prepare an individual so well, he can land on a carrier deck, automatically . . . in shock!)

July, 1953, found the Korean War over. I wasn't going to get into "My War" yet! At the end of basic training in the Pensacola area, I had orders to Corpus Christi. Mixed feelings crept into my mind . . . go to Corpus, get my wings, spend my four years of obligated service, then get out and into the more lucrative airline industry, or stay in the service. When a request came out for volunteers to transition into helicopter training, I signed up, figuring I'd do the required two years obligation in choppers rather than the four left if I went to Corpus. Thank God, I wasn't selected. It scares me now to think of all the fun I would have missed if I had done either of the above.

Corpus was great. After instrument training in the old SNB, I lucked out and got into prop-driven F6Fs instead of the old TBM. Man, did we have a tiger by the tail, the first take-off in that old Hellcat. I had seen this airplane in the movies where John Wayne had even flown it in combat. Heaven had arrived!

Corpus also reintroduced me to rodeo and a wonderful family named Irby. While at Corpus, a fellow Utahn, NavCad Dee Calderwood, and I decided to try our hand at bareback riding. There was a little local rodeo arena called the Horseshoe Inn on the outskirts of town. After church one Sunday, we went out to the arena and signed up. As we were preparing ourselves for the event, an older gentleman, probably feeling we needed help, came up and offered same. Although Dee had ridden a few barebacks before, Mr. Kirby showed us the proper way to put on our newly purchased spurs and a few other well-placed hints about riding a bucking horse. This chance meeting brought me into one of the finest families on this great earth. I drew a big old white horse that just threw me all over his back. I had yet to put one spur to his neck, but I rode him. Dee rode his too.

The next day I was scheduled for a hop in the F6F. I had never ridden a bucking horse before in my life and ol' white horse sure put a hurtin' on my inside groin muscles. That morning, I strapped on my parachute, walked out to my assigned aircraft, and found out I was unable to raise my leg high enough to reach the first step. The Hellcat was somewhat difficult to mount, having a rather high first step. Pressing a button high up on the side of the airplane released a stirrup-like step just behind the wing. As you faced the aircraft, there were hand-holds in the fuselage to assist the "mountee" in pulling up onto the step. After pulling oneself up to a standing position, you simply took a long step sideways onto the wing, walked up to the open canopy, whose threshold was a little high, stepped over onto the seat, and settled the bod down into the cockpit. No easy chore when you had a parachute strapped to your rump.

I tried everything to get my foot up on that step without success. Finally I asked my plane captain to let me hold onto him to keep from falling over backward, and then he could lift my leg and foot up to the step. It worked, but I couldn't reach the handholds to pull myself up. The plane captain put his shoulder under my rump and pushed me up on the step. The move over to the wing was painful, but with his help I managed to put my knees on the wing, only then I found I couldn't stand up. I crawled up the wing on hands and knees and pulled myself to a standing position next to the cockpit. After much discussion as I sat on the canopy rail, the plane captain first lifted one leg into the cockpit then the other. Once inside, no problem–I could work the rudders and brakes. My only problem was being unable to spread my legs apart. The plane captain said, "Sir, can you fly this thing, and are you sure you want to?" I reassured him, thanked him for his help, and cranked her up. God bless all plane captains!

The hop was uneventful. When I returned, my guy was waiting and he helped me out of the cockpit. I sat down on the wing, slid on my butt to the edge, and he then jumped down and lifted me off the wing onto the ground. Unbeknownst to me, some instructors saw me mounting the bird. They were waiting for me as I approached the hangar. I received a pretty good tongue-in-cheek grilling. Luckily, they were Marine instructors and just laughed it off with the recommendation that rodeo was probably not my forte.

Here again was another incentive to go Marine. Fly with most Navy instructors, you flew the hop and came home. Fly with a Marine instructor, you flew the hop, and on the way home, we usually got into tail-chases, acrobatics, dog fights and strafing runs on King Ranch cows. I'll never forget one hop. After our scheduled gunnery syllabus flight was completed, our Marine instructor put us in a tail chase (four F-6s in line, nose to tail, and as close to each other as we dared). He was Lead (radio call signs were Lead, Two, Three, and Four), I was number Four (tail-end Charlie), with two students, Two and Three, between myself and the instructor. During one maneuver, Lead rolled into a high-speed dive, pulled up into a vertical climb, and then performed a slow roll. During the slow roll, I lost sight of Two and Three but still had the instructor. They had completely disappeared! I immediately retarded throttle, dropping back to see what had transpired, mostly to stay clear of Two and Three. Suddenly the two of them reappeared in front of me in perfect tail chase formation. We dove, did another slow roll in the opposite direction, and as we leveled out we heard the transmission, "Lead, this is Two, I am now number Three. Three is now in my old position as number Two." Somehow, and I'll never know, even though I had the best seat in the house, the two of them had switched positions. The instructor simply said, "Way to go, guys, that's a down for the hop. Stay in your present positions, we're making a run on a pickup truck I just spotted . . . hang on! Back in the briefing room, we couldn't figure out how the incident happened, had a good laugh about the whole thing, and Two and Three didn't receive a down for their airmanship.

The next weekend, Dee and I signed up again at the rodeo. I drew the best bucking horse in the string and was riding the hair off him . . . until he made a slight turn to the right and—yep—dropped me right on my head! Too late. I was bitten by the rodeo bug.

We cadets had to live aboard the base, but my weekends were spent with the Irbys. Dee had found a girlfriend. I had found my second family. Mr. Irby owned two mares, a black and a bay. Shirley, his daughter, and I went riding around the countryside and at the rodeo on weekends. I learned my girlfriend was married to a Navy pilot who was on an extended carrier cruise. Now, a good Mormon boy just couldn't take advantage of a situation like that. Shirley never afforded me the opportunity to cut in, either, so our relationship was kept at the best buddies level. We all went to the rodeo every weekend—Mr. and Mrs. Irby, Shirley, and her little baby, Candy.

Before one rodeo, I found out the clown had decided to quit rodeo-in'. I talked him into letting me borrow his clown outfit and arranged to clown the rodeo. Mr. Irby bought me some greasepaint sticks, picked a clown face out of a magazine, and made me up. I used that "face" during my entire rodeo clown career.

I didn't know what the heck I was doing, but the crowd liked my jokes, and I even fought one bull. I loved it! There's something about standing in front of a raging bull, knowing he's fixin' to put a terrible hurtin' on your bod, and when he charges, being able to duck around him and get away. Doesn't beat sex . . . but it's sure next in line.

Red Skelton, Betty Petras and the Author

Author – Rodeo Clown – Lloyd Jack

Chapter 3

Yankee Team

I was a member of VMCJ-1 (May, 1964) stationed at Iwakuni, Japan. Unbeknownst to us, the situation in Laos was heating up. The aircraft carrier USS *Kitty Hawk* (Kitty) was positioned at the entrance to the Gulf of Tonkin and was authorized to begin an aerial reconnaissance and show-of-force operation into Laos called Yankee Team. On May 21, the Kitty launched two RF-8A Crusader photo reconnaissance planes for the Plain of Jars (Plaines de Jarres) into the heart of Laos. Over Xieng Khouang, one of the Crusaders, flown by Lt. Chuck Klusmann, USN, was hit by ground fire, setting a wing on fire. Chuck and his wingman climbed to 40,000 feet where the fire went out, either from lack of oxygen or depletion of the wing fuel, and both landed safely back aboard the Kitty. Seems the carrier needed some more photo birds. The next day, seven RF-8As were headed for the Kitty, and five of these were our Marine photo birds. I was about to get my first nibble at war.

Little thought was being given to or information was available about the rapidly escalating events in the Tonkin Gulf back at VMCJ-1, when an early morning call came from CINCPAC to augment Yankee Team operations with five pilots and aircraft. That morning I reported for work rather early. Major Swede Thuesen, our XO, was already in the office and called me in, saying "We've got to send five pilots/birds to the Philippines on a top-secret mission. The first five pilots coming into work are going. You're the first guy here. Get in your flight suit, make out a flight log to Cubi Point, and include air-refueling over Okinawa. We'll have your radio frequencies before you take off and will take care of your flight plan with the tower. Call us when you land at Cubi for further orders. Hurry, your ground crews are already preflighting your aircraft."

I remember a real quick pilot's brief in the Ready Room and on the way out to the aircraft with the other pilots. The five, Capt. Russ French, 1st Lt. John Sledge, Deej Kiely, Sandy Alexander, and me (a Captain at the time) fired up our RF-8A photo birds and headed for Cubi. Apparently, some of the guys packed some clothes and shaving gear. Not me. I was making out the flight logs. All I managed to take along was one pair of skivvies, one flight suit, my wallet, and one battered VMCJ baseball cap. I lived in that flight suit for the next two months. I did, however, manage to get it laundered occasionally. (For those who worry about dirty underwear, I washed out the skivvies every night.) There was some trouble with Sledge's plane, so he was late getting off the deck, but he arrived at Cubi later that afternoon.

Upon arrival, the Navy was waiting for us. We were told to prepare for a flight out to the USS *Kitty Hawk* CVA-63, but they only had room for four birds. I called the Squadron, as briefed, and was told to stay at Cubi to act as liaison until my bird was needed. Russ, Deej, Sledge, and Sandy flew out to the Kitty the next morning. Their flight included the Photo Det (Detachment) from the USS *Bon Homme Richard* (Bonnie Dick). Damn! I was going to miss something big! Finally (May 25), I received the call, cranked up my bird, and went aboard the Kitty. The Navy flew off some F-4 Phantoms and F-8 Crusader fighters to make room for the photo birds. Seems like someone mentioned there were 14 photo birds on the Kitty. The Kitt's Photo Det, the Det from the Bonnie Dick, our photo birds, and a couple of Navy RA-3Bs had also flown out to the Kitty from the Philippines. Overkill . . . but this seemed to be a max effort.

Right after I came aboard, someone told me one of our F-8s had been rammed into a piece of equipment—a bulkhead (wall) or something—by the ship's aircraft handling crew. When CAG (Carrier Air Group Commander) came to inspect the damage, he stated, "Well, that's okay, it's a Marine bird and not one of ours." This statement really set the tone for our treatment by some of the senior people aboard ship and their feelings toward Marine aviators. A couple of days later, a Navy plane captain accidently discharged a fire extinguisher into the cockpit of one of our aircraft.

Navy Regs (Regulations) stipulate that a pilot cannot fly aboard a carrier unless he has "qualified" within the last six months. I hadn't Car-Qualed (Carrier Qualified) since back in CONUS (US of A) in May, 1963—almost a whole year. The other guys hadn't been aboard for over six months. However, none of us had any trouble landing while aboard ship. We were gooood drivers!

Just to show you what most of our pilots were like:

On a deployment to NAHA, Okinawa, one night while in the lower bar of the NAHA O'Club (flight suit on) run by the Navy, our bunch decided to demonstrate the Marine Corps expertise in flame throwing and eating glass. Russ, Rod Bown, Deej, Pierre Goller, John Sledge, and probably Jim Green were participants. (Jim was our superior glass eater and never missed a glass eating party. He seemed to savor that gum blood.) Sledge loved to bump heads with anyone who was of a like persuasion, and that's how the party started out. (He had a hard head and never lost one of these encounters, although I'm sure he really took a hurtin' a few times!)

They soon got around to flame throwing. The proper procedure for "spectacularity" was to take a mouthful of lighter fluid and then, on a prearranged signal, spit it out into the air towards the ceiling. Someone, preferably a buddy, would light the stuff with a cigarette lighter as it came out of your mouth. A mini flame thrower! A large, African American Air Force pilot happened to take an interest in the festivities and soon wanted to take a turn. After several drinks were bought to elevate his courage and pump his adrenaline up sufficiently, he took the plunge. Just as he was filling up with fluid, Sledge said, jokingly, "I'm from Alabama and I'll light the match!" The guy promptly started laughing and spit lighter fluid all over the floor. (They didn't light it off.) However, a few minutes later, Rob Bown pumped a mouthful out about ten feet onto the wall just above and behind the jukebox. It caught the wallpaper on fire and created a rather nice effect. They were all laughing so hard they didn't even think about putting it out, until someone suggested they all line up and see if they could pee high enough to extinguish the flames. One of the Navy bartenders finally saved the day, and probably the club, when he came over with a pitcher of water and extinguished the blaze.

They settled down then and went to eating glass. Russ had introduced them to the wonders of Brandy Alexanders, and as each guy finished his drink, he was goaded into simply taking the glass edge between his teeth, biting off a chunk, grinding it down on his teeth, chewing it into little pieces, and swallowing the agglomerate. The bar was a large round one, and someone came up with the idea of foot racing around the top. They had conducted several races and were getting to the Kentucky Derby, when the management got a little tired of their escapades and threw them out. The party proceeded to the BOQ and the last thing anyone remembered was Deej, buck-ass naked, chasing someone out the door and around the building with a fire extinguisher. These were the kind of guys I had been flying with for almost a year, and, believe me, it was a fun time to be alive.

After I came aboard, I realized the Kitt's pilots were getting all of the missions. Being senior Marine aboard, I was asked to attend all the Squadron Commander meetings. At one of the first meetings, I suggested a ladder system be incorporated into the pilot priority selection: the last pilot having flown in-country would be placed at the bottom of the rung, with everyone else moving up a notch. I had greased this with the other Photo Det people, so we kinda had a majority. After much deliberation, I'm sure, CAG agreed.

Many missions were flown between May 21 and June 6. Each mission consisted of two photo birds, Lead (Flight Leader) and his wingman. Very few got into the Plains due to the foul WX (weather). Russ rapidly worked to the top of the Ladder. When we came aboard, the Navy folks promulgated a policy whereby no Marine would fly a mission into Laos until we'd gone in with a Navy Flight Leader—a craw-binding idea to a hard-charging Marine aviator. When he got back from his first hop, Russ was rather upset with Navy tactics and navigation. Lead had taken them into North Vietnam, a no-no, and would have pressed on deeper if Russ hadn't "squared him away." Russ had been the first Marine aviator to fly a combat mission into Laos and probably North Vietnam.

Lt. Chuck Klusmann, really a great guy, got hit by ground fire again a couple of days later. Russ and I sat down with him for a serious talk about tactics. We asked how he was flying his hops. He stated, "Just as we've always done—2500 feet and 320K." We suggested this was *war*—not like flying back in the States. (The consensus of our Marine photo pilots was to go down to treetop level, balls to the wall, then get under and through any AAA. Get in quick and get out!)

The next scheduled mission, I sat in on a senior Navy Staff Officer's photo preflight brief. (He was a Staff Officer but had permission to fly with the Kitty Photo Det.) The plan was elaborate, except it included several dead-reckoning turns that were surely going to get him lost. (Although we heard the Spooks (CIA) had set up a portable TACAN station in Laos, we weren't provided the required frequencies to utilize them.) The only navigation aid in the RF-8A was a TACAN, thus all missions to Laos were strictly dead reckoning. I suggested same and offered an easy alternative: fly to Savannakhet, up the Mekong River to the big bend east of Vientiane, thence a heading to the small prominent airfield in the Plains, and start from there, a known and identifiable location. He didn't take my advice, but when he came back, I sat in on his debrief with CAG. He told CAG about the horrible weather, getting lost, etc., and then said, "CAG, I believe the only way to get into the Plains is to fly to Savannakhet, up the Mekong, to the big bend in the river" . . . repeating my words almost word for word.

I finally got my first in-country hop with my Navy Lead (a

Navy Lieutenant). Most of our missions were simple road recces (Reconnaissance) over the main road from the North Vietnamese border around Ban Ban, west through Xieng Khouang, then on to Luang Prabang. During our preflight briefing, I discussed my theory about going in low and getting into the Plains due to cloud cover. As we approached the bend in the Mekong, we faced a solid overcast up to about 8,000 feet; however, it looked like we might get under the stuff. I mentioned that to Lead and suggested we go under the stuff to our target area. He disagreed, saying he thought we might get in a little further north. We couldn't.

While steaming around above the overcast, we spotted another aircraft. It looked like an ol' Army U-6 Beaver, probably Air American (the covert CIA airline in Laos) or an enemy transport. I suggested we go take a look-see, but as we headed in his direction, the rascal turned into us—a threatening maneuver or a bluff. Lead broke off the fly-by, saying it looked like a friendly. We headed for home, and as we passed over the bend in the Mekong, I suggested we make an attempt to get under the overcast to no avail.

That night, Lead and I were standing on one of the catwalks up front. He confided in me the fact he was really tired of the Navy, the long family separations, the time at sea, the flying. He said when he got back to the states, he was seriously thinking of getting out into the civilian world. I knew he was rather depressed, but I felt, deep down, the guy didn't really enjoy the potentially dangerous missions we were flying. I readily agreed that he should get out. After his short Yankee Team sting, he went back to the States. The Navy promoted him to Lieutenant Commander, I'm sure enticing him to stay in the Navy, then sent him back to Vietnam a year later. He was shot down on a photo recce up North. He was a good friend . . . and probably should have been a happy, live civilian right now.

Incidentally, as we passed over Danang, I looked to see if the runway was completed in case a guy had to use it as a divert field.

I couldn't really tell from the height we were flying. I asked people aboard ship, but never received an answer. There were no divert fields listed on the flight info screen in any of the Ready Rooms. I knew it was open to CDR Stockdale after the Maddox incident because he took his wounded-duck wingman to Danang. I believe the reason we weren't told about it being in an "up" status, was that the place was highly classified at the time.

While aboard the Kitty, we were given our "Vietnam Brief." The Navy had flown in an individual who was "craggy-old," wore washed-out khakis, had oak leaf clusters on his shoulders, and was introduced to us as having been in Indochina since WWII. He was an expert on search and rescue in Vietnam and on "the bush" (jungle) over which we would be flying our missions. He lost us when he said that our best bet was not to bail out but ditch (a jet—on land?) and stay close to the wreckage. He was entertaining, though, and told us a story about going out into the bush to find the wounded or dead body of an American serviceman. When they found him, a tiger had been chewing on the poor guy (had gone for the tummy and innards first). They couldn't determine if he was killed in action (KIA), so ol' Bush put in his report "EIA" (Eaten In Action). The guy was a piece of work.

I went solo on my next mission to visually count the NVA (North Vietnamese Army) junk-sampan (ship) traffic plying off the coast from Vinh south to Danang, a non-combat hop. Sandy and Deej had gone out earlier on the same mission and had received AAA fire from Tiger Island. (This little island, just off the coast east of the DMZ and controlled by the North Vietnamese, was a favorite place to dump unexpended ordnance for birds flying back from incomplete hops later in the war.) Neither pilot had observed the AAA and only found out about it after returning to the carrier. A Navy/Vietnamese ship sailing off Tiger Island reported the firing. (Navy didn't even give them credit for a combat hop—more like, "Yeah, you guys really screwed up.")

My hop was uneventful, except for the fact there were

thousands of sails out in the Gulf. I went in low going north. Most of them looked like junks to me. I knew the Navy would never believe the high numbers, and with no way to visually count them, as I came back south, I just simply turned on my cameras and let them do the job. Never did hear the total count from the photo-interps (photo interpreters), but I realized the NVN had a substantial capacity for transporting supplies to the South.

Then came the masterpiece. Up until this point in time, no viable photos had been taken of our objective areas. Photo recon had been greatly hampered due to inclement weather and aircraft getting shot up. Deej launched with his Navy Lead (June 2). Sandy and another Navy pilot were standbys in case one of Lead's flights went down. Sandy was last man on the totem pole. Right after takeoff, Deej's Lead lost a generator and had to abort. The Navy standby was going down the front elevator as Lead stated his emergency, and with Sandy sitting right behind the cat (catapult), they pulled him up and launched him. The two Marines joined up, flew into Laos, received AAA and ground fire, and returned with the best photos taken so far in Laos. Man, were we proud of them! Just after they launched, I was in OPS and was told when CAG was informed that two Marine lst Lieutenants had been launched. He was fit to be tied! I know our guys were sharp kids, both top-notch drivers, and when the folks in OPS looked at me for my reaction, I simply told them not to worry, our guys knew what they were doing and could handle it. We had just blown the Navy's restrictive policy about sending a Marine in-country on the wing of a Navy bird. The photo interpreters were sure tickled. They finally had some good photos to look at. Some pictures! This was our highlight film! Screw you, Navy! (And although both pilots had seen flack creeping up on their tails before taking evasive action, the Navy didn't seem to believe them.)

Now don't get me wrong, we were really treated well by our Navy counterparts. We respected their efforts and recognized them as the true professionals they were. Most of the personnel aboard the Kitty were great folks, but the senior officers who had requested our assistance in the first place really seemed to look down on our capabilities. Looked like all they wanted were our airplanes. Maybe somewhere down the line they felt we had been forced down their throats, and they didn't want us aboard, or maybe they were just getting even.

I remember being at NAS Atsugi, Japan, back in 1957. The Atsugi O'Club was managed by the Navy. At that time, the Navy gents were allowed to bring their families to Japan. We Marines went wifeless during our 13-month overseas tours. It boiled down to the fact that the Navy wives, who frequently visited the bar at the club, started complaining about the boisterous, foul-mouthed, fanny pinching Marines. (Marines are well-known for letting off steam in their O'Clubs . . . any O'Club.) The Navy brass complained, and word was passed to tone it down when there were Navy wives in the bar. Now we Marines have a greeting for people we like. If a person dear to our hearts, say a former squadron mate we hadn't seen for a few months, enters the club, someone shouts, "Hey, let's say hello to Jim!" . . . and all present blare out in unison, "Hello, you asshole!" At the packed "Navy Wife" bar a couple of nights later, when some young (nameless) Marine hollered out, "Hey, let's say hello to the Navy wives" . . . every Marine in attendance answered in unison, "Hello, you assholes!"

There was hell to pay the next day. Our Commanding General stipulated that such behavior would not be tolerated and, in the future, if any complaint was reported, the Senior Marine Officer Present would be held accountable. For the next two weeks, the Senior Officer Present at the bar was a lowly, unknown 2nd Lieutenant. No one wanted to accept that dire responsibility. About a week later, finally realizing all that good Marine money was going downtown instead of into their coffers, the club relaxed the rules and invited us back.

On the 2nd of June, the USS *Constellation* (Connie) was

coming on station to replace the Kitty. We were scheduled to fly over to the Connie and took off on the same launch as Chuck Klusmann's scheduled mission into Laos. He was shot down NW of the Plains. Later we heard a chopper went in to retrieve him. The gooks had set up a helo-trap (Helicopter Trap) around Chuck, and as the chopper let down to retrieve him, the gooks opened up on the bird. (The pilot took a round in the leg.) Sensing the danger, Kluz waved the chopper off and headed for the bush. He was immediately captured and later escaped. Message traffic didn't confirm it, but it sounded like the "spooks" (CIA) sent indigenous people in to get him out. Where did the rescue chopper come from? While the incident was unfolding, we landed aboard the Connie.

On the 7th of June, the first armed-recon "retaliation" missions were scheduled. Ours, the two from the Bonnie Dick (now on the Connie), and the Connie's photo birds were the only photo aircraft available. We were briefed to meet armed F-8s launching from the Kitty over Danang, run our now-escorted mission, and if we "guinea pigs" were fired upon (or hit), the straight 8s were to retaliate.

A morning flight was launched: four F-8s from the Kitty and the photobird from the Connie, led by Commander Wray, a gooooood driver. I believe they received ground fire, had one aircraft hit, zoomed up through a solid cloud cover to 40,000 feet, performed a hairy in-and-out-of-the-clouds air refuel at that altitude, and, thanks to Commander Wray's professionalism, all landed safely back aboard. The flight, however, was unsuccessful either at retaliating or obtaining any photo coverage. The photo pilot on the mission landed back on the Connie at the same time we were being launched, so we were unable to profit from the flight's experience.

Russ was now on the top rung of the ladder and was scheduled for the afternoon hop. I was standby. As always, we were briefed by ADM Guest (Commander Carrier Task Group) and his staff. Four F-8s from the *Kitty Hawk* and Russ were to rendezvous over Danang and go. The F-8s knew not where—just follow the recce bird. Our target was one of opportunity—a road recce between the airstrip in the Plains and then east to Ban Ban. If we took fire, the straight 8s were to retaliate on the guns being fired at us or on anything that moved in the area. Retaliation was stressed. We were going to teach the enemy a lesson: don't shoot at our photo birds. How the enemy could distinguish a photo bird from any other jet was beyond me, but, in the background, there seemed to be a push to get us deeper into the North Vietnam/Laos conflict.

Only three F-8 fighters from the *Kitty Hawk* made it. Russ lost his radio receiver before arriving over Danang and couldn't join up visually. I was about to shut down when I got a call to launch. Airborne, I switched to the flight frequency and could hear my flight rendezvousing with a refueling tanker over Savannakhet. I gave them a call, reported in as their recce bird, and headed for their position. Weather was CAVU (Clear Air-Visibility Unlimited) over Danang and most of the way to Savannakhet, where I joined-up, topped off with fuel, took the lead, noted the side numbers of the flight (call signs were the same as side numbers), got a radio check with each member of the flight, and headed for the bend in the Mekong.

We hit the bend in the Mekong, dove under an overcast, and, "balls to the wall," picked up the heading for the Plains. I could see our entrance was going to be blocked by a large rain squall dead ahead, so I turned right, picked up a heading to Ban Ban, and, spotting two dirt roads ahead, turned on my cameras. We were really streaming along under about a 400-foot overcast—the tops of surrounding hills were covered by clouds—when I got a call from someone in the flight saying I was starting to walk away from them and could I give them a couple of inches (take off a little power to allow them to catch up). I eased back on the throttle and glanced back to see if the flight was staying with me. I saw fuel streaming out of CH-7's right wing. (I had no way of knowing the pilot's name, but later, back at the ship, found out that CDR Doyle Lynn's side number was Charlie Hotel 7.) I hit my mike to ask if he was in trouble and received the reply . . . "CH-7 is streaming fuel . . .looks like he's been hit . . . he just punched out!" I called saying, "You guys stay with him. I'll go High Cap and get some help."

I zoomed up quickly through the overcast, breaking out at about 7,000 feet. Then it hit me: who do I call now? These missions were still being classified Top Secret. I couldn't go out on Guard frequency and tell the world we had just lost a Navy bird over Laos! What had Chuck done? How had the rescue choppers found him? (We had never been briefed on Chuck's rescue. His flight had been controlled from the Kitty Hawk.) We hadn't been briefed on SAR procedures nor been given any frequencies to call in case of a downed aircraft. I switched to Connie and then Kitty freqs, but couldn't raise them. I tried to call my flight back, but no luck. Talk about a void.

Finally, I went out on Guard frequency. "Any aircraft in the vicinity of Vientiane receiving me on Guard, this is Romeo Mike-14, please switch to 251.6 (my flight's frequency)."

In about 30 seconds, I got a call . . . "RM-14, I'm reading you. What's the problem?"

"I've just lost a wingman south of the Plains. Can you help me?" "Where did he go down?"

I described our route of flight and told him if he stayed under the overcast, he should see the smoke from the crash.

"Roger, I'm on my way, but that's a pretty hot area."

I told him there were two F-8s circling the pilot and that I was going to altitude to call my people. I assumed I had just talked to a spook (CIA/Air American) chopper.

I climbed to 20,000 feet but couldn't raise a soul, flew down to Savannakhet, raised the ship and reported CH-7's plight. They replied that refueling aircraft were being launched and requested the downed pilot's name, which I didn't have. I didn't seem to be able to convince them of the latter. Insisting that the call sign of the downed aircraft was CH-7 off the Kitty, I told them I was going to fly back up north to see if I could assist in any way. By now the "bend" was socked in. I flew north, found a couple of small holes in the overcast, went back under, but couldn't find enough maneuvering room to identify any landmarks. I tried, flew back to the "bend," but try as I might, couldn't get under the foul weather, nor could I find the usual black plume of smoke that accompanies a jet aircraft crash.

Later, I received a call from the Kitty, saying refueling aircraft were over the area west of Danang on such-and-such a TACAN radial/distance. When down to bingo fuel, I headed for the refueling rendezvous. All I could think of was . . . here I've lost a wingman in Laos . . . and I can't do a damn thing to help the poor guy. Shit.

Upon arrival over the designated spot, I had no radio communication with anyone and no refuelers. Realizing they were tuned into Connie's TACAN, I switched to same and headed a little north, finally picking up the refuelers (two A-4 Skyhawks). Their Lead hand-signaled me to switch to his frequency. After establishing radio contact, I took on fuel from both aircraft and told Lead I was headed back in, would try to get under the cloud layer, and locate the pilot with my cameras. He said, "Wait one," and came back with, "Ship is trying to locate you, they want you back aboard! . . . they need questions answered." Lordamighty, I had just taken on a load of fuel to go back in to find my wingman, and they wanted me home. I inquired into the status of my other two wingmen. Lead told me they had refueled and were returning to the Kitty. Nothing to do but go home.

When I landed back aboard the Connie, I was ushered up to the Bridge to join "all the Brass in the world." While describing my route of flight, some black-shoe Captain said, "You mean you were lost and don't know where the downed bird is?" He swore, turned on a heel, and stomped off. (How embarrassing.) I finally got across to these people that, no, I couldn't tell them exactly, but the mark on my map was in the ballpark, and, as soon as my film was developed, we'd know for sure. An hour later when I sat down with the photo interpreters, we were able to match up some landmarks and pinpoint almost the exact location. I had been in the air four hours. It seemed like a couple.

Early the next morning, I was called down to the Operations Center. Upon arrival, I saw three or four Brass and G-2 (Intelligence) types sitting there, and off to one side a pilot I recognized from the Kitty. I was told the Lieutenant was one of the pilots in my flight . . . get your heads together and tell us where Lynn was shot down. We readily agreed on the spot. The Navy then grilled us for about 15 minutes on the conduct of the flight. The F-8 driver said, "As Lynn punched out, he noticed a tin-roofed looking building off to his left. He swung over to get his sights on it and fired two rockets, didn't see the hits, and when he swung around in a 360-degree turn to take a look, he couldn't reacquire the target." No two ways about it, these interrogators were looking for some blown-up targets to show we retaliated. The Brass practically browbeat the guy trying to get him to admit that he had hit a target. I really felt sorry for him. They just acted really pissed-off that we hadn't accomplished something. (LBJ was probably looking right down their throats.)

Finally someone decided the debrief was over, and they left the two of us alone in the room for a few minutes. Our conversation revolved around the reason for the grilling.

"Hey, what in the hell is going on? Did I hit a pagoda, a school, a hospital? What are these guys looking for?"

I ended the conversation, saying, "Man, they were really giving you the fifth degree. Seems they were looking for a little retaliation. Guess they were just teed that you never rocketed Laos…and sunk it!"

About that time, the Brass came in and took the poor guy away. From the hostile environment of the meeting, I figured they might pull his wings, kick his butt off the plank, or find a yardarm to stretch a little neck, and I might go with him. I don't know what happened, but I never saw him again. (I found out later they stuffed him into the duty helo and flew him to the Kitty, now on its way to Cubi Point and then the USA.) Needless to say, the hop was pretty unsuccessful. (To this day, I have never found out the exact particulars of the events after Lynn punched out, what action my other wingmen took, how Commander Lynn was rescued, who rescued him, or how he was returned to the Kitty.)

Scuttlebutt reported Commander Lynn rescued the next day. They might never have found him if one of our guys hadn't given him some of the pistol flares we had been issued back at Iwakuni. (They were not in the Navy inventory yet.) These flares came in a packet containing two pencil ejectors about the size of a felt-tip marking pen and several cylinder-shaped flares. To operate, you simply cocked the ejector, screwed a small flare cylinder into the top of the ejector, released a button on the side, and poof—a red flare shot up 100 to 200 feet into the air. One of our Marines, probably Russ, gave Lynn an ejector and two flares just before we left the Kitty.

When the choppers went in to locate Lynn, they couldn't find him through the jungle canopy. Lynn could hear the choppers, fired a flare, choppers saw it, and came to a hover right over him. The jungle canopy was too high for the cable and seat to reach him, so he grabbed some logs and built a little platform to stand on, but it was still three to four feet above his reach. Lynn climbed up into a tree, but the sling was too light to penetrate the jungle canopy, and the chopper crew couldn't get it down to him. The chopper pilot, concentrating so heavily on the rescue, settled into the treetop, over-stressed an engine pulling out, and had to limp back to his base (wherever that was). From his vantage point in the tree, Lynn spotted what looked like an opening in the foliage where the second chopper pilot could pick him up and signaled the pilot to meet him there.

The second chopper went to the wrong area, soon realized it, went searching again, and, as a result of another flare, located Lynn and pulled one relieved Commander out of Laos. Lynn somehow made it back aboard the Kitty. They all went home (USA). (Lynn came out a few months later and on his first hop got shot down by a SAM (Surface to Air Missile) up around Vinh . . . two hops, two "shot downs" . . . snake bit.) I have no idea where the rescue choppers came from, but I suspected CIA/Air America involvement. (We now know that Air America was flying out of many "Lima Sites" in Laos.) All missions into Laos were canceled!

The next morning, the Connie's Photo Det decided to hold an "Atta-Boy" meeting in the Ready Room. They let a lot of their enlisted men into this briefing, but I didn't see any of our kids attending (we had about 20 to 30 enlisted troops aboard by now). During the Atta-Boy, I realized our troops had been kept in the dark about our secretive missions into Laos, and if the Navy thought it appropriate to tell their people, I could certainly tell our kids.

As I exited the Ready Room, I saw the whole gang standing in the passageway. I queried them on why they hadn't attended the briefing. They were told there wasn't any more room. I wanted to make this brief special—highly patriotic—so I figured the Hangar Deck, seated around one of our birds, would be the ideal spot. I asked them to follow me out there. Now, Marines are not known for their familiarity with aircraft carriers. The maze of ladders and mystifying passageways can be rather mind boggling to the uninitiated. With the whole bunch in tow, I proceeded to go up and down ladders, around and through passageways, back through these same passageways, until I finally ended up in a dead end with about twenty chuckling Marines in trail. One of them said, "Captain, do you want me to find the Hangar Deck for you?" With egg all over my face . . . yeah . . . we were there in about 30 seconds.

Our enlisted men had been doing a fabulous job keeping our birds flying. (Navy pilots were tickled to death when they were assigned to fly one of our birds on a mission.) Our guys worked long hours, and, I'm sorry to say, they were sometimes billeted in terrible quarters and got little comfortable sleep, usually ending up sleeping in cockpits, under and on top of wings, and even in jet air intakes. They did one heck of a job aboard the carrier . . . God bless 'em all.

During the lull of in-country missions, the Navy flew a couple of night launches. After the first one, I attended a Squadron Commander's meeting. They were discussing the lighting on the carrier deck. We Marines were not night qualified, so I sat there rather disinterested. Opinions were requested as they went from one CO to the next, each in turn giving his thoughts on the best way to light up the carrier deck. I had never seen a lighted deck from an aircraft on approach, but from observing night OPS from "Vulture's Row," the whole process seemed a might hairy.

After we flew over to the Connie, the four of us were given the VIP quarters, just through the bulkhead from Admiral Guest's quarters. One night, Sandy, 1st Lt. Roger Williams, and Russ set out to get all the sailors drunk. (Deej and John Sledge had returned to Iwakuni, and Rog was their replacement.) Russ gave up the ghost and rolled into his sack about midnight. When Sandy and Rog fell in the door about 0300, Russ and I were rudely awakened. While trying to get the two quieted down and in bed, Sandy promptly marched over to the "Admiral Wall," repeatedly banged his drunken fist on it and screamed, "Admiral, you old fart, git your ass over here and join the party." Russ went white as a sheet. A quick discussion revolved around jumping overboard or staying on the ship with the distinct possibility of being thrown in the clink, with bread and water for the rest of our lives. We all finally decided it might be better to just get in bed and hide under the covers!

We never heard one single word about this incident, but later that day, the grapevine rapidly spread the word about a surprise booze inspection in the officers' quarters today, so if you got any, get rid of it. I guess we really ruined it for the Navy. I'm told there were a lot of bottles floating off behind the ol' Connie that day.

The period from June 11-18 saw all in-country flights come to a halt. Commander Stockdale's (later of the Maddox incident and POW fame) squadron of fighters had flown over from the USS *Ticonderoga* (Tico) to supplement the Connie's fighter squadrons. We practiced some armed recce flights around the ship: four F-8 fighters to go in for flack suppression, four A-4s to drop bombs on enemy positions, four F-8s to mop up with rockets, and then old photo recce in to take pictures (one in right after another). The whole thing took about five minutes. Our new CAG wanted that photo bird in right on the heels of the last gunships. Russ and I discussed this and agreed that all you were going to get were pictures of smoke and debris (good photos come after a pilot determines when the target is clear). On my first hop, Commander Stockdale led the hop and was flying as Lead in the first flight of F-8s. The commander's flight made their first run, but as the other aircraft attacked, he called, saying he wasn't satisfied with his four-plane strike, and would I mind if he went in again in front of me. By the time they got out of the way, I was about 30 seconds late. Boy, did CAG chew a piece of my tail. Here's a guy wanting us to put on an air show for the AAA gunners in Laos! We had quite a discussion about smoke and debris. Stockdale finally broke it up, and, seeing the reasoning behind our argument, talked CAG into letting the photo bird pick his time over target.

Most of these practice flights were flown within sight of the carrier, usually ending up just off the port beam. A lot of the ship's company came out on deck to watch these practice flights. I was a real proponent of the "go in low, balls to the wall, and get out quick" tactics, but I began to notice how slow the photo birds appeared. We all flew them in afterburner or at max power, and I thought they would go by like zzoommm; it was more like chug, chug, chug. The gist was they looked extremely vulnerable to enemy fire. I know we lost a few birds later on up North following this theory, but I still felt these tactics were proper. (Figured the gooks had been lucky to pop Lynn . . . they were probably shooting at me or the flight in general.)

During this lull, the carrier would drop off a sled and tow it along behind for target practice. The flight deck would fill up with people to watch the Navy practice bombs and rockets on the thing. It was really fun to watch our intrepid friends, and there were some real professionals amongst them. The 14th of June saw a restart of Yankee Team operations; however, restrictions to avoid further losses insisted on a "minimum risk approach." Reconnaissance aircraft had to fly over 10,000 feet, above the range of most communist anti-aircraft guns, unless we needed low-level photographic coverage of a target. Hell, this was going to take all the fun out of flying for us Marines. Our birds were configured with cameras that cried for low-altitude photography. Needless to say, the quality of aerial photos diminished considerably.

Around the 10th of July, the Connie was about to go offline to be relieved by the USS *Ticonderoga* (Tico). I was designated to fly over to the Tico to brief the oncoming Photo Det. The Navy strapped me into an RA-3B, in one of those seats that face the rear of the aircraft, a rather precarious way to take off and land on a carrier for a Marine aviator; however, the flight was uneventful. I quickly settled into the briefings and the local movie theater. I stayed the 11th and saw Commander Stockdale's F-8s come back aboard the Tico. (The joke around the Tico was that Commander Stockdale's pilots had become so accustomed to flying off the larger Connie, they might have trouble even landing on the small, wooden-decked Tico, but no problem—they came aboard like true professionals.)

On the 12th, I woke up about 0430, walked up on deck, and, low and behold, sitting out there on the briny was my old friend, the Connie. I noticed the Tico's SAR helicopter getting ready to fire up, and, having no further tasks aboard the Tico, I ran over to the chopper and asked the driver where he was headed. His answer was, "Just up around the ship." I asked him if he would run me over to the Connie. He replied, "Jump aboard!" I took a seat in the cargo area. As we were in the landing approach to the Connie, I glanced out a side window and couldn't believe my eyes. There, standing at attention, were Admiral Guest, his staff, and an honor guard, all in dress whites, surrounding a rather long, red carpet.

I thought the gathering was some kind of official heroic medal presentation or a reception for a VIP, and the chopper driver was mistakenly landing right in the middle of the ceremony. Before I could unstrap and jump out of the chopper, a full commander from the admiral's staff stuck his head in the open doorway and, rather puzzled, asked, "Where's the captain?" Trying to be helpful, I answered, "Sir, I'm the only passenger aboard, but I am a Marine Corps captain." The commander went a little white, then red, mentioned a deity, spun on his heels, and walked off toward the formation. It didn't take me long to dive over the side of the ship onto a catwalk and disappear deep into its bowels. I knew somebody was going to catch hell and hoped it wouldn't be me. Believe me, I tippy-toed around that carrier, trying to stay out of the admiral's path.

I found out later that Admiral Guest was just sitting down to breakfast in his cabin, in his skivvies, when the chopper pilot called, about halfway over to the Connie. He stated he was on his way over from the Tico with either "THE" or "A" captain from the Tico and requested permission to land! I guess the Connie folks decided it was "the," not "a," and there was one hell of a scramble to assemble all the participants and one red carpet.

I managed to duck Admiral Guest for about a day and a half, but I knew the confrontation was eventually going to happen. Finally, as I rounded a corner into a passageway leading to our cabin, almost to the security the door would provide once inside, the Admiral and his entourage appeared around the corner. I braced the bulkhead, expecting some sort of dressing down, but as the admiral passed by, he simply smiled, shook his head from side to side, and said, "You damn Marines!" and walked away chuckling. I knew he was referring to Sandy's little 0300 caper on "drunk night."

We grew to really like and admire Admiral Guest. He sat in on every brief and debrief when we flew into Laos from the Connie and really treated our pilots super. I saw him stop many times to engage the ship's enlisted men in conversation. When Russ and I were briefing or being briefed by the Admiral's staff for our flight on the 7th, he repeated the famous statement, "Never have so many been led by so few!"

After cross-decking from carrier to carrier for two months, our original five pilots were soon replaced by other VMCJ pilots. I flew one of our sick birds to Cubi Point for repairs, and after spending some lovely liberty in Olongapo, hopped a ride back to Iwakuni in a transport, my war finished. We were scheduled to return to CONUS a short time later. My appetite had been whetted! I wanted back in!

While at Olongapo, 1st Lt. Roger Williams, one of our replacement pilots, and I went out on liberty. As we were walking down the street, little kids kept coming up to us, pawing at us, and begging for whatever. I acted real mad and told them to get the hell away from us, even slapping at a couple of them. Rog was flabbergasted and made a remark about how mean I had treated them. I warned him that they were only doing it to get close enough to grab a wallet out of a pocket or a wristwatch off an arm. He still felt I had played the ugly American, but soon changed his mind when all of a sudden one of the bold ones grabbed his Rolex and tried to jerk it off his arm.

A day or two later, the Connie came into port. I had thoroughly briefed most of the Connie bachelor officers on the delights liberty had to offer in Olongapo—the best bars, how to get there, which girls were hot, the ground rules. They had been at sea for better than a month and deserved Olongapo! I warned all the bar girls to be ready and in general prepared the town for the Connie crew. What a ball! The first night they hit town, about twenty of them came bounding up the stairs into the designated bar, eyes bulging out of their sockets, with saliva drooling from the corners of their mouths. Man, the girls almost ran out the back door! In my estimation, they picked the ugliest girls in the room (the Navy always displayed a lack of class). Many fell in love that night and stayed that way, right up to the time they went back on station the 5th of August.

All of our original Det were recommended for DFCs by the Kitty Hawk. Up the line, the awards were knocked down to Air Medals, and then to:

"The Board is of the opinion that the flights herein listed fall more appropriately under the category for which the Strike/Flight system was created and are not so exceptional as to merit a personal decoration." I agreed, except Sandy and Deej deserved a DFC for their photo hop into the Plains. They really stuck their necks out, taking fire in the area where Chuck Klusmann had been hit three times by enemy fire and shot down. Many DFCs were awarded during the conflict for flights under lesser circumstances.

Chapter 4

The Interim

While on the Kitty Hawk and Connie, I heard rumors about Air America hiring pilots to fly in Laos (à la the Flying Tigers during WWII). The one I liked the best had Air America pilots flying T-28 bombing strikes out of Vientiane, Laos, and somewhere in Thailand. The local requirement— having a "native" pilot aboard the aircraft— was usually met by strapping a Thai or Lao cook into the back seat. These rumors were confirmed when I talked to one of the Spad (A-1H Sky Raider) drivers while on the Connie. The Spad pilot had flown into Vientiane to refuel, had actually talked to some of the Air American pilots, and observed the T-28s sitting on the ramp. The Chief Operations Officer was a former Spad (probably still was), and a Marine buddy of mine. Rumors said he had stashed away over $100,000 in the two years he had been flying for Air America. I had just lost a wingman over Laos and sure wanted to throw some metal in retaliation. I certainly couldn't do that from a photo bird.

I left the Yankee Station carriers for Japan with the desire and address to visit Air America in Washington D.C. My orders after leaving Japan were to Cherry Point, North Carolina, so I went a little out of my way just to visit D.C. I was pleasantly surprised when I walked into their office. Plush—this had to be a first class outfit. After stating my business, I was ushered into a beautifully decorated office and introduced to my interviewer. I stated my purpose: flying for them in T-28s in Southeast Asia. He took me back somewhat by saying, "We don't have any T-28s in the area you are talking about." Although I confronted him with the Spad driver's information, he wouldn't budge on his position. The gist of the conversation steered toward flying transport aircraft or helicopters, pay scales (not as high as the rumors depicted, but substantial) and living conditions, etc. I left the office with a bunch of papers, an application, and a job offer if I wanted it. I didn't. Flying transports and helicopters was not my way of fighting a war. I didn't know, however, they were flying O-1 Bird Dogs.

While stationed at Cherry Point, I was assigned to VMCJ-2. On May 3, 1965, a beautiful Sunday morning, I received a call from a fellow officer telling me to pack my bag. We were headed for the Dominican Republic via Roosevelt Roads, Puerto Rico. The DomRep had heated up and they wanted some photo work. Immediately, visions went racing through my mind: this might be My War! Six hours later we landed at Rosey Roads, Puerto Rico, and when our photo aircraft maintenance troops arrived later that night, we were in business. Our Det (detachment) consisted of two pilots, two F-8U Crusader photo birds, two small photo processing vans, and about twenty photo aircraft maintenance troops. We started flying into DomRep territory the next day.

We didn't know it at the time, but the Air Force had sent a photo detachment from Shaw AFB, South Carolina, to work out of Ramey AFB, Puerto Rico. On our first hop, we overflew Ramey and were rather surprised to see about a dozen RF-101 photo birds sitting on the mat. That night, we received a Frag Order tasking us with every-other-day coverage of targets in the DomRep. The Ramey boys flew on alternate days. We blanched when told to make 32 copies of every photo for distribution to every command, right up through the Joint Chiefs of Staff . . . all out of our little processing vans. One of the copies was to be delivered to the Shaw Boys at Ramey.

We Captains took it on ourselves to say to hell with that. (The request would have severely taxed our capability back at Cherry Point.) But talk about assets—we flew over to Ramey to sneak a peak at their photo operation. Lord what a surprise. We were ushered up to the Det Officer's office, a full Bird Colonel. His Executive Officer/Photo Officer was a Lt. Colonel. They had flown to Ramey the previous night with six birds, several photo processing vans, and enough people to support an army. After he read the Frag Order, the

Bird immediately sent for more assets and ended up with six more birds and a total of twenty one vans. Man, they took over a whole hangar, briefing rooms all paneled off, with equipment all over the place. This was a huge operation. They agreed with our "Captain" thinking. The Bird sent out a dispatch requesting some slack. We got it: send only the pertinent stuff, only 12 copies, which kept most of our photo troops up all night long to process and print the damn film. I suspect the Air Force was sent to secretly map the whole island, including Haiti. We flew a lot of road recce and airfield photo, but the photo intelligence was negligible. Hell, this was banana stuff. On one of my road recces, I was filming with my forward firing camera down the main north-south road between Puerto Plata and Santo Domingo. As I passed over what was the largest town on the route, for gosh sakes, I saw horses tied up to hitching rails in the middle of town. Our troops got a big kick out of that. Never thinking we would be taken up on our invitation to visit us at Rosey Roads, we extended same to the Bird and his XO. Much to our surprise, one morning they dropped in. Lordamighty, here we are sitting in a little cubbyhole the Navy had lent us, our briefing room and office combined, our two Crusaders sitting on the ramp, two photo vans sitting out on the grass about a quarter of a mile away, a 1st Lieutenant in charge, and, if we hadn't been Marines, we'd have had no respect. The XO spent a considerable amount of time with our Photo Officer and on departure made the statement, "I don't see how you can produce the output you do in this environment, but I'll have to admit, you put out a better product than we do. Your photo quality is outstanding!" We could have kissed him. It was a real morale thing for our troops. Our commitment lasted the better part of a month, and we flew back to Cherry Point.

My tour of duty at Cherry Point was interesting. The Navy finally got even with me for Olongapo. I Night Carrier-Qualed aboard the USS *Independence*, off Jacksonville, Florida, in the RF-8A Crusader. Our bunch flew aboard the carrier on March 30, 1965, and were extensively briefed on Night Ops. On the 2nd of April, we were launched during the day to accomplish our Day Quals. That night I was the only VMCJ pilot to make the schedule. Right after I launched into the dark, dark night, the carrier diverted me to NAS Jacksonville (JAX) because of a fouled deck. Several other Marine aircraft were also diverted, and we were rather surprised to see one of our fighter jocks make a beautiful landing at JAX in the dirt, right between the two, well-lit runways.

We were told to relax during the next day and were given a time to report over the carrier that night. Radar picked me up right out of JAX, brought me right down the chute, told me to drop my hook, and "call the ball" when I had it in sight. (I didn't understand the latter because my first two landings were supposed to be touch and gos.) Never saw the blasted ball . . . or the ship! I was given a "wave off" to the right. The next thing I noticed as I stabilized upwind—the three red lights outlining the carrier's mastheads were off my left wing, slowly moving aft. I felt much better knowing the ship was really there! My next pass was fine until just before touch down . . . then everything turned to worms. I was on airspeed, on glide slope, and all of a sudden the ball started going low. I immediately added power, saw a slight red ball on the mirror, made a big correction, and landed. Later I found out I had "taxied" up to the number one wire! Not a comfortable feeling at all (everything short of a "taxi up" was a ramp strike). I was fully prepared to be taxied up to the catapult for another launch but was again surprised when they taxied me up to the front of the ship, parked me, and told me to shut down. The ship was having more trouble with the arresting gear and was suspending any further night ops. I was a little shook, however, when I was told later that the ship was having problems with the mirror and the platform Landing Signal Officer (LSO) hadn't even seen me until just before touchdown. Further, our own Marine LSO hadn't even been told I was in the landing pattern.

Later that night, the ship came back online, I launched, shot two touch and gos, two traps, and was again diverted to JAX. The next night I flew out to the ship, where, for some unknown reason and against my protests, I shot two more touch and gos and than the required additional four traps. As I pulled out of the arresting gear on my last trap, the ship told me to pull up on the Cat for another shot. (Night landings aboard a carrier were always tense, but a Night Cat Shot was unbelievable: zip to 150 knots in three seconds, no visual horizon, relying strictly on instruments . . . it took awhile to get your eyeballs uncaged!) En route I queried the reason behind this unseemly maneuver, because my last trap had been my sixth. Their reply indicated the ship needed additional landings to meet some kinda quota. Well, this night thing was by now pretty comfortable, so I let them launch me two more times, screaming, "Oh, ssshhhiiitttttt!" all the way down the Cat.

Some of the fun, fun Det hops we had out of Cherry Point were our electronic warfare classified missions around Cuba in ol' Willy the Whale (EF-10B). We simply flew four birds one morning to Key West, took another hop around the north or south side of the island, listening for and plotting any electronic emissions beaming out of Cuba (specifically to see if Castro had been issued any more missiles since the Kennedy/Kruschev confrontation). After spending a good liberty night in Key West, the next morning we would fly around Cuba and land in Montego Bay, Jamaica, for "Crew's Rest." Lord, someone had certainly set up a beautiful deal for our Squadron. We took our ground crews along, had rip-roaring times, and seldom rested.

Liberty started as soon as the birds were tied down and everyone was sequestered in one of the nicer hotels on the beach. We partied that night, the next day and night, and the next morning leisurely went out to the airport, fired up the birds, flew the north or south side of Cuba, and landed at NAS Guantanamo Bay, Cuba (Gitmo). Here was the real crew rest. There wasn't much to do but sleep at Gitmo. The next day found us flying back around Cuba to Key

West, then reluctantly back home to Cherry Point.

These missions proved fruitful for some time after the Cuban Missile Crisis but had slacked off immeasurably. As soon as Castro's boys saw us headed for his island, they shut down most of their electronic gear. There was always one exception to this procedure. The Cuban radar operators gradually became familiar with our flight paths around the island, so occasionally we would terminate our leg on the north side of the island, make a nice little turn to the north, and wind up in Nassau in the Bahamas. I guess the "enemy" relaxed and thought we were headed home. After our normal two-day "Crew Rest" in Nassau, we fired up the birds and headed straight south for Cuba. Man, every electronic appliance on the whole island came buzzing up on our screens with a wealth of information, and, let me tell you, if the liberty in Montego Bay was great, the liberty in Nassau was incredible!

I kept hounding my detailer up at Marine Headquarters to send my butt back to Vietnam but was told I had to wait in line, as there were many Marines ahead of me. I finally received orders to the Naval War College in Newport, Rhode Island. I was disappointed, but I had a great time at the "gentlemen's college." Especially spectacular was the classified library that was heaven to a guy who loved history. New England opened up new avenues for touring—Niagara Falls, Toronto, Ottawa, Quebec, the Boston Red Sox and Celtics, the fall foliage changing colors, and, of course, the great seafood.

At the War College, we students were following the war in Vietnam every day, so I was itching to get back into the thing. I had contacted my detailer at Headquarters expressing my desire to be sent to Vietnam as soon as possible and expected orders right out of the War College. Not so. In June, 1966, I received orders back to Cherry Point. While working in the Group Commander's office as an Assistant Operations Officer, I checked out in the old, ancient C-117 (Gooney Bird). One weekend, we flew the Marine Corps Band and Drill Team down to Dallas, Texas, to perform in the Cotton Bowl during a Dallas

Cowboy football game. It also happened to be the weekend of the Oklahoma Texas game. Man, does that town turn up the activity level for that one! The Marines performed at many functions during the three days of revelry, but my "impressionable moment" was at the downtown parade.

I was standing on the sidewalk just as the parade was starting. Traditionally, the band and drill team lead off the big parade. The crowd started to animate as the band came down the road, still about 50 yards away. The man in front of me had a little boy about three years old at his side. As the "Marine Corps" approached, he lifted the young one up onto his shoulders and exclaimed, "Look son, here come the Marines, look how straight and proud they march, and listen to the music . . . it's the Marine Corps coming! The Marine Corps is coming!" The whole crowd echoed his sentiments, which made me one proud Marine.

About a week later, I lucked out and got another hop to Dallas. I was sitting in the O'Club, ready to hit on any unsuspecting female that was of like persuasion, when an old friend from my Atsugi, Japan, tour in 1957 walked in the door with a beauty on his arm. This guy was one wild character, a heck of a pilot, and was best known for bringing a busload of deaf Japanese girls through the main gate into the Atsugi Officer's Quarters for a night of gaiety. We got to talking about old times; his beauty could have cared less. She finally spied an acquaintance at the bar, left our table, ended up dancing a couple of times with her friend, and then came back to our table. Rick didn't bat an eye or seem to notice her absence. (Rick is a pseudonym for reasons unbeknownst to me or him.) A little later when the guy came over to the table to ask her to dance, Rick said, "Yes, she can dance with you . . . but take her with you, I don't want anything more to do with her, and if you give me any guff, I'll knock the shit out of both of you." Needless to say, they both departed the area in a hurry. Rick was now working for the Secret Service in the Dallas area, and when

I mentioned the Marine participation in the Cotton Bowl, he said he had seen both games that weekend without buying a ticket. Seems he would go up to a ticket taker, flash his badge, and say, "See that guy that just went through the gate, the one in the blue blazer wearing yellow pants? I'm tailing him for the Secret Service." The keeper of the gate usually let him through the turnstile.

Air America somehow entered the conversation, and, to my surprise, Rick had been one of their employees. This really peaked my curiosity. He had flown in Laos, had "crashed" three fixed-wing passenger airplanes, all while attempting landings at the same airstrip. It seems this field was on a high mountain peak, was very short, and had a sheer cliff on the approach end.

This cliff dropped off straight down about 100 feet onto a plateau. Seems the first two accidents occurred when Rick got caught in a downdraft, wiped out his gear on the edge of the cliff, and ended up in a belly landing on the dirt strip. Luckily, he and his passengers survived. On the last wreck, carrying several passengers, he had been sucked down below the ledge by a wind shear, flew smack dab into the cliff face, fell down to the plateau, and was the only one to walk away from the crash. He didn't have a scratch on his bod. He was fired.

I heard through the grapevine that VMCJ-2 was formulating plans to TRANSPAC six EA-6A aircraft from Cherry Point to Danang, Vietnam. Either J-2 didn't have enough pilots qualified in the bird or they couldn't find anyone who wanted the job, so they needed one more pilot. By then, January of 1967, I had re-fammed in the EA-6A and had about thirty-five hours in the aircraft. The Group CO, knowing my desire to get to Vietnam, called me into his office and said, "I can't finagle orders for you to Vietnam, but I can give you the next best thing, a TRANSPAC." Man, this was a chance at another adventure, so I grabbed the offer. During the first week of January, I flew eleven more hours to get a little more qualified in the bird, but the one refueling hop they scheduled for me was canceled due to non-availability of a refueling tanker (KC-130). I was told, "No sweat. You can hook up on the leg to El Toro." I believed them because the EA-6A refueling probe sat on the nose of the aircraft and appeared to be a cinch to hook into the basket.

On the 13th of January, we started our "ball buster" to Danang—six EA-6As and a straight A-6A—and I mean "ball buster" first leg: Cherry Point to El Toro, with an air refueling over Oklahoma City (six hours and ten minutes). Now, as any jet jockey knows, the pilot can stand up, move around, and relax in a transport aircraft, so six hours is not that long, but strapped into a seat where there is nothing to do but sit and fly the bird, the old butt and body can get stiff and sore. Air refueling was a snap: I hit the basket on the first stab. After a day's rest, we briefed on the El Toro Hawaii trip with the KC-130's refueling people. The CO was going to lead the flight and was rather concerned that I had only stabbed the basket once. He firmly stated, "If everyone in the flight is not hooked up well before the point of no return, the whole gaggle is returning to El Toro," and he was looking straight at me. Now, the EA-6A has all kinds of navigational equipment on board, everything a guy needs to get from here to there. Only one problem: none of it worked. We hit El Toro with all systems down, except for a couple of TACANs. The A-6A with the most sophisticated navigational system in the Marine Corps wasn't faring much better. Lead's TACAN was down, but he bummed one off another airplane. Mine!

The second leg of the trip, six hours and forty-five minutes, found us at Kanehoe Bay, Hawaii. We had some difficulty spotting the KC-130s at the refueling point, overran them, and during the left-hand turn to establish our position behind the refuelers, Lead pulled off throttle to slow down and dropped his nose. I was flying the number two slot on the left wing of our leader and quickly started to stall out. I dropped the nose to gain airspeed and flew under the flight. It took me several minutes to get into position on my tanker, and on my first plug in, I missed. Man, how embarrassing! I tried again and missed. Ridiculous! I was still cussing my ineptitude when I hit the basket on my next try.

Every once in awhile when plugging in, a situation that pops up that can become extremely hazardous, especially when the only thing under you for miles and miles is the Pacific Ocean. As you stab the basket, the hose automatically pulls back into the tanker (called "response"). This is done to allow for the closure rate between your aircraft and the tanker. If this slack is not taken up properly, the hose starts to snap up and down violently, causing a tremendous strain on your refueling probe. This usually results in a bent probe, incapable of receiving fuel. Yes, friends, as I plugged in, I got a discerning whip from the hose. It became so violent I had to back out. I hooked up again with the same results. I asked the tanker to check the response on their aircraft. They checked it out, I hooked up again, same result. Suddenly I received a call from the lead KC-130, "Aren't you hooked up yet?" My answer was, "Not yet, we're having trouble with the response on my tanker." His next rather curt transmission was, "If you are not hooked up within five minutes, we are headed back to El Toro!" I hooked up to my tanker two more times with the same result. I guess the lead tanker finally checked with my tanker (on another frequency) to determine which of us was the problem. I didn't hear the conversation. Finally, he admitted my tanker was having response problems and sent me over to another tanker, which happened to have the last free basket. I'd had a lot of practice recently, so plugging in was no sweat, the response worked fine, and I topped off with fuel. Then that son-of-a-gun kept us hooked up for another twelve minutes, and it's not much fun hooked up to that basket at 200 knots, loaded with fuel, and just hanging in thin air. We were certainly relieved to cut loose.

The next leg was a six-hour job to Wake Island. Lead still had a TACAN, and with Wake's radar, we found that little ol' island. We always had a day layover at each stop on our route, so we absorbed the history of Wake Island while in the O'Club. Preparing for takeoff the next day, we noticed several Air Force O-2 Skymasters taking off for Guam. They were replacing all their O-1 Bird Dogs in Vietnam, had stripped the O-2s down, added internal fuel tanks, and were flying them across the briny. I wondered what luck they were having with their navigation systems. Compared to the air time these little bird drivers encountered, I didn't feel so sorry for our guys on our six-hour hops. After what seemed like fifty or sixty of these little birds took off, two at a time, we received clearance and flew to Guam without incident.

The next leg into Cubi Point, Philippines, was uneventful. The last two legs were three-and-a-half hour trips and were rather comfortable. Although I was the senior officer on the Transpac, I relegated authority to the VMCJ-2 flight leader, so during the whole trip I didn't have any responsibilities except as a pilot—a nice relaxing situation. After arrival, the Brass decided our aircraft would remain at Cubi, the six EA-6As at Danang would be flown out of country to Cubi, the switch made, and we would fly their planes back to Cherry Point. There were only thirteen of these birds in the world. They were sorely needed in Vietnam, and the Brass didn't want twelve of them sitting on the deck at Danang. I wasn't even going to see Vietnam! I couldn't say the same for Olongapo. We stayed at Cubi four nights and renewing old acquaintances in Olongapo was pure bliss.

Our trip home went via Midway, instead of Wake, which was interesting. Liberty in Hawaii was fine as frog hair. I even attended a Rodeo. (There are some excellent cowboys on the islands.) The hop to El Toro was smooth as silk, and a day later we headed home, but in a roundabout way. We had to skirt some bad weather out of El Toro, filing for NAS Pensacola, Florida. Home the next day. The whole trip was fun. We saw a lot of ocean, but it's not a trip I would volunteer for again.

RF-8A Crusader Photo Jet

The Big Birds

EA-6A Intruder *RF-4B Phantom* *EF-10B Skynight*

A Phantom Reposing on Rubber Tires
Waiting Out a Typhoon

(Note tire/wheel tied to wing tank)
The Morley Far East Transport Company

Air Fueling from KC-130

Chapter 5

Danang

It didn't take long after the Transpac to get orders to Vietnam. I could hardly wait. My insignificant arrival in Danang occurred on March 14, 1968. I caught a ride over to MAG 11 to check in and was assigned to VMCJ-1, which tickled heck out of me. VMCJ-1 was flying the RF-4B (Phantom) photo bird, the old EF-10B (called F3D, F-10, or, more affectionately, Willy the Whale), and the EA-6A (Intruder). The latter two were the electronic warfare birds. I only had ten hours in the F-4 and hadn't flown it since 1966, so most of my time was spent flying the electronic planes.

The missions for the F-10 were simple. You strapped into a Willy somewhere between 1800 in the evening and 0600 the next morning, flew up to the DMZ, set up a racetrack pattern between there and Dong Hoi, spent your required two hours on station, and flew back to Danang. The right seaters (Electronic Warfare Officers, EWOs, or, as we called them, R/Os Radar Operators) did all the work. I considered myself simply a truck driver, although stinking weather could make the trip rather eventful.

Designed as a carrier, a jet night fighter, during Korea, ol' Willy the Whale had a side-by-side cockpit configuration and excellent fuel capacity, afforded good vision, and had an emergency escape chute behind the pilot and R/O. The right side was packed with passive receivers, analyzers, recorders, jammers, chaff dispensers, and other special mission electronics.

"These valuable machines performed brilliant work during the Quemoy and Cuba crisis, and from early in the Vietnam involvement flew more hours per aircraft than any other squadrons, bar none . . . without getting any publicity. It was the pioneer EW/ECM platform in that theater, and continued in use after arrival of the USAF EB-66s and even after arrival of the much bigger, and more sophisticated, EA-6A." (Gunston, Bill. *Aircraft of the Vietnam War.* Patrick Stephens Ltd, 1987.)

The escape chute was something else again. It had a small, pressure-sealed door situated on the rear wall of the cockpit. The door could be hand operated, or, in an emergency, blown open by pulling a handle on the console panel between the pilot and R/O. The handle also blew open the floor of the chute at the bottom of the aircraft. The side-by-siders then had to work themselves out of their seats somehow, R/O first, grab a bar at the top of the open door, and pull through and slide down a chute, somewhat akin to a child's slippery slide.

I was introduced to the ol' F3D at MCAS Cherry Point, fresh out of the training command. She wasn't a fashion plate or one of our hottest, but I fell in love with her. One day someone decided to give us new guys practice on bail-out procedures. A Willy was pulled out onto the hot concrete ramp, the escape door opened, a couple of mats thrown under the hole in the bottom, and we were briefed on the proper procedures. The most important point: R/O goes first, hits the mat, and with great dispatch, rolls out of the way of the pilot now coming down the chute. This was a hurry-up, don't wait procedure. A couple of teams came down the chute without any problems except for the hard landing. Somehow the next team was a real mismatch—the smallest R/O and the largest pilot in the Squadron. About halfway down the chute, the R/O got hung up on his shoulder harness, stopped dead in the chute, shut his eyes, and immediately got stomped into a little ball. Somehow they both bottomed out of Willy at the same time. During my 18-year affiliation with Willy, I never saw anyone practice the procedure again.

One incident with Willy comes to mind. Scenario: The pilot's intercom switch was located on the throttle. Push forward, you transmitted outside the aircraft, thumb the switch backward, you were on intercom and could talk to your R/O. Additionally, in takeoff position, as you thrust the throttle forward, if you forgot to turn the air conditioners all the way down, vapor came storming out the ducts, frosting over the windshield and filling the cockpit with a white mist. Also, we were starting to see women Marines manning the tower as traffic controllers. (It was sure nice to hear a sweet, pleasant voice transmitting from the tower.)

I was taxiing out for takeoff one day just as another Willy was running up for takeoff. The neophyte pilot, startled when the mist surrounded him, hit the mike switch to talk to his R/O . . . the wrong direction . . . and broadcast to the world, "What's going on in here? We've got smoke up the ass!" Not missing a beat, the gal in the tower called, "Aircraft with smoke up the ass, you're on the air!"

The EA-6 mission was a little different. Electronics-wise, the EA-6A was a huge advancement over ol' Willy. We flew her up the coastline to a position twenty miles off Vinh, North Vietnam (twenty miles off the coast was just out of SAM range), and then set up a racetrack pattern in support of the Navy and sometimes Air Force strikes attacking North Vietnam. (When EA-6s weren't available, ol' Willy took over.) I'll leave the electronic details to some Electronic Warfare Officer to write, but, in general, we had the capability of jamming anti-aircraft guns (AAA) and surface-to-air missile (SAM) radars, rendering them harmless against our strike aircraft. (We could also jam a SAM's radar guidance system, making the missile go ballistic.) Many pilots flying up north, who were attacked by SAMs, were surprisingly relieved when a SAM didn't track and destroy their aircraft. Little did they know their butt was probably saved by an EWO in ol' Willy or the EA-6A.

We really had some great, well-trained R/Os. They performed little miracles every day and never received the credit they deserved for their contribution. However, most of our electronic recon missions were highly classified, which rendered a paucity of publicity.

When the Navy birds didn't get launched or were delayed for some reason, we augured around the pattern picking up on enemy electronic emissions. The hops generally were two-and-a-half to three-hour trips, but sometimes, when your relief didn't show up, you had to refuel from a Navy tanker and could stay on station for another two hours. These were really satisfying hops . . . not much excitement, but you felt a real contribution to the war effort.

Many of these electronic warfare missions were flown at night. In fact, I didn't have a day hop out of Danang for two months, this being when I jumped into the right seat with an ol' hand and got my Area Fam (Familiarization) hop up to Dong Ha, Khe Sanh, and our divert field at Ubon, Thailand.

I was itching to get in a bird, but policy stated you had to wait three days to get the required briefings and walk-arounds. During one of these briefs, we were told about youngsters, women, and old men assisting the VC. They cautioned us to watch out for local folks coming up to you, appearing to be friendly, then tossing a grenade at your feet or having an armed grenade in a can, tossing it into your jeep as you drove by. Worst of all, if you went to bed with some cute gal, she might take a sharp knife to those private things we all cherish. (Rumor had it that, a year or so earlier, liberty in the town of Danang had been curtailed after a taxi pulled up to the main gate sentry box and, before taking off at high speed, rolled out the decapitated head of a M/Sgt. who had been in town.)

A day later, a buddy asked me if I would like to take a jeep ride up to Freedom Hill, the location of the big PX. To get there, you had to go out the back gate, through a little town called Dog Patch (somewhat VC-controlled), then travel four or five miles up a dirt road to Freedom Hill. Going through Dog Patch, you wouldn't even know there was a war going on. Not too many people waved at us, though. As we approached Dog Patch on the way back, I spotted an old man standing along side the road talking to a lady. He was holding what appeared to be a Hills Brothers coffee can cradled in his left arm. Recalling the briefing I heard the day before, I mentioned the old man to my buddy, who didn't seem too impressed. We agreed, however, if the old guy made some overt-looking move, we would floorboard the jeep and get the hell past him. About twenty yards out, the guy shifted the can up to his right shoulder (like a shot-putter holding the shot). That was enough for us. He jammed down on the accelerator, and I prepared to swat back any incoming can. The old boy didn't even seem to see us passing. We both had a rather embarrassed look on our faces . . . and a pretty good laugh at our folly.

Danang was continually harassed at night by the VC. In fact, when taking off to the south, frequent tracer rounds could be seen shooting up at our aircraft. Takeoffs to the north involved the VC Navy. These little guys would grab a fishing pole, jump into a rowboat, row out into Danang Bay (just off the runway), pack a rifle or an automatic weapon for protection, and then shoot at our birds just lifting off the runway.

Many stateside critics, ungrounded in the aspects of the simplicity of war, could not fathom the fact that a place as large as Danang could be hit with rockets or mortars. They never stopped to think that some little VC, working in his ville all day, might sneak out one night, place some aiming stakes up, and the next night or next week, sneak out and hide a base plate next to the aiming stakes; the next night, he'd haul a couple of rockets/mortars out to his site, put all the hardware together, aim in and fire four rounds, close shop, and return everything to its original hiding place. Impossible to stop unless neighbors turned him in.

One story goes: A Marine pilot in the landing pattern at Danang observed some rocket impacts on the airfield, saw the position the rockets were being fired from, sucked up his gear, flew over the site, and took it under fire, effectively eliminating the threat. The next day he was kicked out of Vietnam for firing on a target without prior permission. (Oh, the rules of war.)

We Marines were located on the west side of the north-south

runways, and the Air Force toiled on the east side. I must say, however, it was a little discouraging to see the Air Force going into their air-conditioned trailers while we retired to our little dust covered, screens-for-windows, southeast Asian huts.

One night after soaking up a few suds at the O'Club, some of the VMCJ gang started up a volleyball game out in back of our hut. When I arrived, having a reputation for honesty, I was asked to referee. I hustled back to the O'Club and bought three fifths of vodka. I took my position at the net and cited the now-existing rules: anyone really screwing up, losing a dumb point, questioning a referee's call, or even looking askance at the ref had to take a big swallow out of the vodka bottle, and if the error was stupid enough, he had to remove an article of clothing (a volleyball strip tease). The rules were accepted with great merriment by the rather tipsy crowd, and the games began. The proceedings just got drunker and drunker and nakeder and nakeder. As the games wore on, there were a few casual remarks about the long and short of things, the flopping arounds, the circumcised and uncircumcised. We drug off the falling-down drunks and readily replaced them with eager volunteers. By the time I left the shindig, there were only three or four guys on each side, the rest having been carried to their huts, or they had become bodies strewn about the court. Three of the bodies were still laying on the court at 0600 the next morning. Picture this . . . twelve Marine officers, naked as blue jays except for their black flight boots, jumping up and down . . . you got the picture!

Soon after my arrival at Danang, someone asked me if I had gotten into the three-man lift yet. It didn't take a Harvard graduate to figure out this was some kind of Marine Corps prank, so I went looking for a friend. I found one, so didn't get trapped. But others weren't always as lucky.

I had never been a participant but had observed the Navy's Junior Officer's Watch or Sea Bat shenanigans. The Junior Officer's

Watch was pulled on many a Marine aviator . . . all unfamiliar with carriers. Now, all Marines stand duty watches, so when the Navy aviators came up with the Night Duty Watch List, it was not taken lightly by the Marine jocks. The gullible ones were taken to positions around the ship, signed into the log book, and told their relief would appear in four hours. It's been said that some carriers had as many as eight Marines standing the watch at one time, and after the watch stander's four-hour stretch, they went looking for the guy who didn't show up to relieve them.

Our enlisted troops weren't forgotten either. Their nemesis was the Mail Buoy Watch. Once given a lookout station on the carrier, they were given binoculars, told to keep a lookout for the Mail Buoy, and report when it was in sight. It seems ships at sea have certain locations where they pick up their mail. The mail ship sails along until it sights the mail buoy, drops off the mail for the carrier, and when the carrier streams by, the mail buoy watch informs the ship's personnel that the mail buoy is in sight. It was also a four-hour watch.

The sea bat thing revolved around a large crowd gathering around on the Hangar Deck to see the huge sea bat some sailor had fished out of the water. The uninitiated were told, "Jeez, you ought to go into the center of the crowd and see this big sea bat some guy had caught. Biggest fish I've ever seen." The big fish was hidden under a tarp, and as the victim leaned over to see the thing, some swabby, usually a Chief, sneaked up behind the poor guy and whacked him on the butt with a large paddle that resembled a baseball bat.

The three-man lift, however, was ingenious and took considerable fanfare and planning to complete. It usually occurred after a USO show or a special occasion in the O'Club when most of those in attendance were in their cups. Some guy would jump up on the stage and proclaim, "I've got a man here who can lift 500 pounds, and here's twenty bucks to back him up!" Well, 500 is a substantial weight for what a normal guy could lift, so the betting started, mainly by the previously initiated. Soon, sizable bets were laid down. The barker would then solicit three volunteers (new guys) from the audience to act as weights. Great concern was taken to get the right-sized guys to make the 500. Finally, with the bets settled, the three selected were placed on the floor and wrapped up toboggan style, and the lifter would approach the weights. Just as he reached for a hold, he would back off, and every soul (100 strong) within throwing distance poured their drinks on the tobogganers! Man, I'll tell you the three came up spitting and gasping for breath. While the build-up progressed, every guy in the bar ordered new drinks. The crowd would then gather upon the scene, and I say "upon" because there were guys hanging from the rafters with pitchers in their free hands. I talked to one of our civilian Tech Reps who had been one of the three. He said it was so bad, he truly thought he was going to drown.

During the end of my tour at Quang Tri, we did a lot of work with 3rd Recon, whose boss, a Lt. Colonel, occasionally came to our O'Club. One night he brought this huge 1st Lieutenant with him. This guy was big, a weight lifter from the old school. The Lt. Colonel took over the stage and, with his arm around ol' Huge, stated, "I have a Recon Marine here who can lift 500 pounds." Bets were placed, three of Huge's fellow Recon officers got on the toboggan, and the scene was perfectly set. Ol' Huge took time out to do a few stretching exercises to get warmed up . . . I mean this guy was really serious! He finally got his hold on the three, gave out this great "aaaaAAARRR" . . . and promptly got drowned. As he came up, sputtering and gasping, he was ready to take on the house. He was one mad Huge. He still couldn't see too well but went grabbin' for the first guy in reach, with fire in his eyes. The Lt. Colonel, a little guy, sensing an aviator massacre, grabbed the huge one by the lapels, backed him up to a wall, and got him quieted down. Thank God, I didn't see any of our guys wanting to take him on.

On one of my first DMZ hops, I was flying up the coast just

off the city of Hue (pronounced whey . . . like curds and), when some other pilot called, asking for my position and altitude. My answer was, "At 12,000 feet over Huooey" (like the duck). He laughed and said, "My God, man. How long have you been in country? It's pronounced Whey, Whey!" I realized my mistake, felt like an idiot, shut up, and didn't talk to him again.

On the bus returning from lunch one day, I saw a big CH-46 helicopter lifting off from the 1st MAW landing pad. He experienced engine failure and auto-rotated to a nice landing, right on top of the Wing Post Office. Looked just like a giant grasshopper seducing a lumber pile.

The Navy support and DMZ hops were fun, but the highlight of my day would be the nightly movie at the club. The movie theater was located on the patio adjacent to the club, had a nice secure roof, but was "open air." Luckily, it was on the side of the club farthest away from the town of Dog Patch. Small arms fire was not uncommon from the town, which was located about one dirt road and a fence from the club (some 25 yards away). Not too many Marines walked between the club and Dog Patch at night. Movie fare included many spaghetti westerns that featured killing upon killing. It was not uncommon to see thirty bad guys on horseback running towards two good guys, the latter firing five shots apiece, and killing all the bad guys. We kept a running killed-in-action (KIA) tally each night, and it would sometimes run into the hundreds.

Actually, we felt pretty safe at Danang. It was very comforting to hear the B-52 strikes going off near the Base, with Puff the Magic Dragon firing its beautiful stream of tracers to the ground or our artillery firing every night. I often wondered what my reaction would be to an incoming rocket. I didn't have long to wait. One morning about 0300, I had just returned from a Navy hop and was catching up on the paperwork at my desk. Suddenly, I heard this screaming whistle and immediately found my body prone on the floor. I hadn't told it to go there, either. I heard four blasts go off right outside the building where I was working but found out later that the rockets had gone off a couple of miles away in Danang City.

The EA-6 was designed to be the optimum in the electronic warfare (EW) business and the most advanced bird of its time. There were only thirteen in the world. We had six at Danang, six at Cherry Point for training, and another one somewhere being used as a test bed. That's why we had to stay twenty miles off the coast of North Vietnam. We couldn't afford to lose a bird. The EW side was designed with a taping technique, allowing us to fly over enemy territory, suck up all the enemy electronic emissions on recording tape, stay on station to provide jamming support for attacking aircraft, fly home, and automatically record every thing on a huge wall map.

Now, like every other automatic thing, nothing can go wrong . . . go wrong . . . go wrong. We could never get the things involved up to speed; they were just too sophisticated for our technicians who hadn't been sufficiently trained to keep the systems in an "up" status. The pilot's side included a search radar system and a navigational computer that would have been a tremendous asset off the coast of NVN. We could never get either system up long enough to be of use. Even civilian tech reps who came direct from the factory couldn't get the things to work. We had to revert to using TACAN radials from one of the destroyers auguring off the coast southeast of Vinh. This was fine during VFR weather, because you could always see the coast. But at night, fighting thunderstorms, and not knowing exactly where the destroyer was, you had a great chance of getting inside the SAM 18-mile ring. We had a couple of our pilots fired upon by SAMs. Call it luck or the skill of their EWOs, they were never hit; however, due to our good jamming work, we were a lucrative target if we happened to slip inside the SAM ring.

Our avionics guys and the tech reps really sweated over the search radar/NAV systems. Finally, one day, one of the tech reps came into my office with a big smile on his face. We had a bird with all systems operational. He had found one little wire, hidden deep down in a black box, that had been wired incorrectly at the factory. They scheduled me to fly the bird that night, so I went over and over the operating procedures. If something went wrong, I didn't want to be the culprit. When signing off the yellow sheet, one of our avionics types told me that the NAV computer might be doubtful, but everyone was confident the search radar would work. Everything checked out before takeoff. As I proceeded up the coast, the radar was working perfectly. God had just turned on the light! The NAV computer even worked until I hit Dong Hoi; then it went down. Even without the NAV, this was going to be a fun hop. Off Vinh, I had a clear radar picture of the coastline and the little islands just offshore. Man, I could put myself exactly over my orbit points. Then, in my first turn south, the radar screen went black. What a disappointment for everyone. We never did get the two systems up and working during my remaining stay in J-1.

One day, after a boring three-hour hop on the DMZ, I came back to my desk and found a note directing me to report to some Colonel up at Wing Headquarters, the purpose being to interview me as the new Wing Electronics Officer. Lord, here I had been busting my balls to get into this war, had been in country less than a month, and someone had nominated my butt to go fly a desk. I was going to make a lot of heel marks on this one. When I met the Colonel, I stated my position, not mincing words. Bless his heart, he told me to return to the Squadron and have them nominate someone who had been in country longer and had a great deal of Electronic Warfare savvy. An EWO, a good friend, ended up with the job. I really felt sorry for him, but he later told me he rather enjoyed the assignment. I sure hated to screw the guy, but I was going to stay in this war.

Sensing the handwriting on the wall and out of the Squadron after six months in NAM, I might be assigned to some desk job. Many were, and I started thinking about matriculating into the O-1 Bird

Dog business. The hops out of VMCJ-1 were challenging but could be rather boring—a truck driver. The O-1 was an environment I could really sink my teeth into, help the grunt on the ground, and see my results. One day when I wasn't scheduled, I went over to Monkey Mountain (Marble Mountain), sat in the O-1, and immediately fell in love with the little bird. An old friend and former squadron mate, Colonel "Big Ed" Reid, a great guy, was the CO of the Helicopter Group that had the O-1 Bird Dogs under his command. He extended an invitation to come over, get checked out in the O-1, and fly with them. I was way too busy to take him up on his kind offer but vowed I'd be in that little bird someday.

Riding the bus to work one morning, we saw one of the huge fuel tanks on our side of the Base on fire. We assumed it had been hit by the VC as a couple of rockets had been fired onto the field that night. Come to find out, one of our troops had fired a night flare into the sky to see if there was enemy movement on the wire in front of his position. The wind caught it during the parachute descent to the ground, and it landed on top of the fuel tank, burning its way through, and setting the fuel on fire. I had always figured that, under like circumstances, the whole thing would erupt in one massive explosion, but that tank took on a slow burn that lasted for several days.

The hops up north certainly produced many human interest stories and some spectacular fireworks. Unless you were in a position to observe the Marine/Navy/Air Force's night attacks up north, you could never realize the tracer rounds and flak flying skyward out of the North Vietnamese weapons. Tremendous! I really take my hat off to the pilots who took on those gunners every day and night. It took some large balls to go into the north, day after day.

We always stayed on station until all the attack aircraft were "feet wet" and headed back to the carriers. One time we heard an aircraft call mayday. He had been shot up but was proceeding back to the ship. We monitored his radio transmissions throughout the ordeal.

He had been shot up too badly to bring the bird back aboard, ran out of fuel, and had to bail out alongside the ship. He was recovered. Another time, we heard and saw an A-4 limping out of NVN toward the coast. The pilot kept losing control of the aircraft, and just about the time he called "feet wet," the whole thing turned to worms. He ejected right on the coastline. We heard the parachuted jock talking to his wingman on his handheld radio. He happily stated the wind was blowing him out over the ocean, where our SAR folks could retrieve him from the sea. His hopes were fortified by his wingman, but his last transmissions indicated that, no, the wind was drifting him toward land. Upon landing, he was quickly captured and probably spent some time at the Hanoi Hilton. God bless our POWs.

One time we were flying a Navy day hop in ol' Willy, which had two 20-mm cannons in the nose. We always carried ammo, and the guns were kept in working order. We were headed directly north toward the tiny island of Hon Matt (located about ten miles off the coast near Vinh). It looked and sounded like all the attack birds were safely feet wet, but suddenly Red Crown, our controlling agency, called stating we had an unidentified bogie directly on our nose. I spotted the bird flying at about 3000-4000 feet in a tight right-hand turn, rolling out over Hon Matt, and then it came full-bore right up at us (we were at 16,000 feet).

We were a rather lucrative target for the enemy, but I couldn't figure why they would risk an attack this far off the coast. I armed the guns just in case. Hell, he could have shot us down with rockets before coming into our gun range, but what the heck, I might get in a lucky shot and be the first guy to shoot down an enemy jet since ol' Willy was used as a night fighter in Korea. I suddenly realized the bird was so big it had to be a Navy RA-5C Vigilante recce bird who had probably just made his BDA (Bomb Damage Assessment) photo runs over NVN, had just updated his NAV computer by flying over Hon Matt, and was now headed for the carrier. However, I wasn't sure, so

I held up my intensity level (some called it the pucker factor) until the bogie zoomed under me without demonstrating any overt action. I pulled a quick 180-degree turn and saw the Vigilante disappear into the wild blue yonder. With tongue in cheek, I called Red Crown and informed him that his bogie had just passed under my position and was now headed for the fleet. He came back with, "We identified the bogie as friendly. I forgot to inform you. Sorry about that."

One day while returning from a Navy hop, I heard a pilot eject from his bird about ten miles north of Dong Hoi. His wingman couldn't raise him on the radio. He orbited for awhile, then departed to refuel. We hung around just in case we could provide some assistance. The wingman came back and continually flew over his downed buddy, trying to raise him on the radio. When the Jolly Green Giants arrived (Sikorsky HH-3E helicopters), they refused to go in to search for the pilot. Man, did this upset the wingman and any other pilot listening in. I myself felt this was a chickens--t crew who were rapidly destroying the fine reputation all pilots had for the Jolly Greens. The monkey came off his back a little when he told the wingman that standing orders specified the birds would not go in if there was no radio contact with the downed pilot. Too many of them had fallen into the ol' helo trap the enemy frequently used to get at the rescue choppers (i.e., they would quickly and quietly surround the downed pilot and then shoot down the rescue chopper when it landed or hovered to extract him from the jungle).

The wingman was still mad as hell, stating he had flown over the pilot now for about 45 minutes, had seen him, and hadn't received any enemy fire. Nothing more was done and darkness set in, negating any rescue attempt. I had the early, early hop the next morning and, as we flew up the coast, picked up the rescue attempt again. The choppers were still refusing to go in. Finally, driven by the incensed wingman, they headed for the downed pilot. Just as the chopper started its approach, automatic weapons opened up, killing a crew member and severely damaging the chopper. The chopper limped away, the NVN picked up the downed aviator, and the rescue attempt went for naught. Blame? Could the chopper have gone in and made the rescue, without taking fire that night? Would the chopper have gone in if the wingman had not insisted? Who knows . . . probably only the little guys on the ground.

Another incident happened that really struck at one's heart. We were orbiting off Vinh in between Navy strikes when we heard radio transmissions involving a flight of two Navy fighters trying to intercept two enemy fighters headed south near Vinh. I had recently read intelligence reports indicating the NVN were flying a couple of interceptors into a field south of Vinh. After landing, the fighters were heli-lifted to a bunker some two or three miles from the field, so we couldn't spot them. When needed, a bird would be towed out from the bunker, picked up by chopper, placed on the end of the runway, fired up, and launched. The intelligence folks figured this venture was in place to sucker our fighters into a trap (i.e., enemy fighters would fly south from their sanctuaries in the Hanoi area). The Navy BARCAP fighters protecting the fleet were sent to intercept. Then, typically before contact, the enemy would not stand and fight but reversed course and moved toward the north. As our fighters tried to outrun the enemy birds to complete the intercept, they had to break off contact just as the enemy flew into their "Hanoi Safety Ring." (Again, the silly rules of engagement.) Intelligence figured an enemy fighter would be placed on the end of the runway and then, as our fighters were concentrating on the two headed North, the enemy pilot would take off, zoom, climb up behind our streaking fighters, and, before they could react, the enemy aircraft would engage our fighters from behind.

A ground intercept controller directed Easy 6 (a Navy flight) and his wingman to start an intercept on two bogies headed south, near Vinh. Easy flight acknowledged the transmission but with a voice about two octaves higher than normal, indicating this guy was extremely nervous or pathetically new at the job. As Easy 6 started his intercept, control indicated the bogies had turned back to the north. The ensuing tail chase was in progress when control reported that Easy 6 now had a bogie at his six o'clock position, climbing rapidly. Shortly thereafter, Easy 6 made the intercept, and the enemy birds split, one going high as the other one dove. Easy 6 called his wingman, stating, "I've got the guy going high, you take the low one." You couldn't hear the sounds of battle, but suddenly, in a descending, almost tearful voice, Easy 6 called: his wingman, Easy 9, had just been shot down. The aircraft had disintegrated with no observed ejection. The intercept was terminated. No report of the six o'clock bogie was forthcoming.

The next day, we anxiously awaited the intelligence reports to see if, in fact, a heli-lifted enemy aircraft had made the attack from behind Easy's flight, confirming the successful enemy tack. We found the dispatch and read it in dismay. Seems Easy 6 went after the high enemy aircraft, lost sight of his bogie, scanned the sky to reengage, spotted a bogie, and shot him down . . . happened to be his own wingman. (The rapidly climbing six o'clock bogie was not mentioned.)

On a Navy hop one night, they didn't show up, so I decided to auger around in case they sent in another strike. Suddenly, my right engine flamed out. We got her started again, no problem; then about two minutes later the left one quit! It restarted, but two flame-outs on the same hop was enough for me. Home—sorry Navy. The rest of the hop was uneventful. After the bird was thoroughly inspected on the ground, it was put back into service. The flame-outs never occurred again and to this day remain a mystery.

One dark, dark night, one of our better young pilots returning from a hop in Willy experienced a double flame-out while letting down to enter the traffic pattern. He satisfactorily restarted both engines, and they immediately flamed out. He obtained another restart, but the rascals flamed out again. His final restart occurred at 400 feet above the water. The rest of the hop was uneventful! Maintenance never found the problem, and the bird stayed in commission.

During our many boring, all-night surveillance flights off the coast near the DMZ, most of our pilots became friendly over the radio with the Radar Ground Controllers (KLONDIKE) located at Dong Ha Airfield. We soon recognized their individual voices and, although against radio procedures, would frequently laugh and joke with them to help relieve the tedium and tension. A great rapport developed. Our call sign was Cotton Picker. One night, when an Air Force aircraft checked in, the transmissions went like this:

"Klondike, this is Cross Bill 12 checking in, over."

"Aircraft calling Klondike, roger, say again your call sign, over."

"Klondike, this is Cross Bill 12, Cross Bill 12, over."

"Aircraft calling Klondike, sorry sir, but I can't understand your call sign, please repeat, over."

Rather irritated, "Klondike, this is CROSS BILL 12, CROSS BILL 12, CROSS BILL 12, OVER!

I couldn't stand the temptation any longer, hit my mike button and said, "Klondike, this is Cotton Picker 26. Try . . . MAD . . . DUCK . . . OVER."

After a slight pause came the reply, "Roger, Cotton Picker, roger, I've got it. CROSS BILL . . . CROSS BILL, thank you, sir!"

Another night, a Viet Cong rocket attack hit Dong Ha Airfield, igniting a fuel and ammo dump. They were still being hit by NVN rocket and artillery rounds. From our vantage point at 20,000 feet, the nighttime visual fireworks were spectacular. As I checked in with the ground controller on the radio, the conversation went like this:

"Klondike, this is Cotton Picker 21 checking in. Man, what's going on down there? The place is sure lit up."

"Cotton Picker 21, roger, we've been under attack all night. I've got an ammo dump cookin' off right next door. I'm the only guy here in the bunker, and I'm under the biggest table I can find. It's rather hectic down here. I hope this bunker holds up!"

The next transmission came from an orbiting Navy aircraft.

"Klondike, this is Navy 31 checking out, looks bad down there, good luck, Buddy . . . and I sure hope your relief speaks English!"

We were always assigned an altitude to fly while working with Klondike to keep hard airplanes from running into each other. Upon arrival at the DMZ, we would clear our altitude with Klondike, who then assumed responsibility for aircraft separation. This altitude assignment was particularly critical when we were flying "in the soup" (inclement weather). One night I checked in at an altitude of 16,000 feet, my area of operation for the next two hours. I was experiencing very unruly weather—rain, lightning, and thunderstorms trying to tear ol' Willy apart. The whole area was bad. After about an hour of fighting this "mung," an Air Force aircraft checked in. The radio transmissions:

"Klondike, this is Air Force 11 checking in at 16,000 feet."

"Air Force 11, this is Klondike. Your assigned altitude will be 20,000 feet tonight."

"Klondike, Air Force 11, we're always assigned 16,000. I was briefed for 16,000 feet. Why are you giving us 20,000?"

"Air Force 11, I have a Marine aircraft operating at 16,000."

Apparently, flying at 16,000 feet and getting rather nervous about a possible midair collision, Air Force 11 came back with, "Is he at my altitude?"

Klondike, probably feeling a little paternalistic about our close working relationship, answered, "No, sir. you are at *his* altitude!"

Frequently we received word from the G2 (intelligence) folks that a VC rocket attack was expected to hit Danang sometime during the night. To save our EA-6s, we sent as many as possible to the USAF base at Ubon, Thailand. We could then send the birds out on a mission the next morning without having to return to Danang. One night, three of our birds ended up at Ubon. We were in the Air Force O'Club, soaking up a few drinks to fortify us for our impending trip to town, when I spotted an old Air Force buddy from our tour at the Navy War College. Murphy by name, he was flying FAC missions out of Ubon in the new Cessna O-2 Skymasters. My talks with him about his FAC missions over the Ho Chi Minh Trail convinced me this was where the action was—the way to fight a war. Our conversation inflamed my desire to transition over to O-1s.

The trip to town was eventful! Outside the gate, we found pedi-cabs waiting to take us to town (tri-wheeled bicycles with a little canopy seat on the back for passengers). I picked out the strongest looking driver and bet the drinks I could beat the rest of them to town about a mile away. I weighed about 200 pounds at the time, so I pleaded for a handicap for my driver. We all agreed I would start down the road, come to a specific spot where I would then stop, look back, and give them the start signal. I didn't stop and signal. Before they became aware of the fact, I was at least one third of the way to town going like a bat. The drivers seemed to be enjoying the race—mine was particularly enthusiastic—and by the time we hit the three-quarters pole, we had a nice lead. Then disaster struck as our chain came off! I jumped out of the cab, told the driver to carry on, and got behind the thing to push. Guess who was last? Needless to say, the bar girls looked just as lovely as the girls back home, and we all had a great time.

Enemy attacks seemed to step up about this time. One night, the three aircraft taking off in front of me were shot at. I could see the tracers. What to do? Hell, if they couldn't hit those guys in front of me, how could they hit me? I went. (So did everyone else—no big deal.) Later that night, as I cleared the runway and was taxiing back to our Squadron area, three rockets went off right between the runways with aircraft all over the place, and not one plane got hit.

One night, one of our EA-6s blew both tires landing at Ubon. The next day, our Maintenance Officer, a great pilot and innovator, Major R.J. Morley, simply tied two mounted tires, one to each wing tank, on ol' Willy and flew to Ubon. Never done before, but it sure worked. This little incident happened before my arrival at Danang, so I was a little surprised when someone came into my office and asked if I would like to fly a tire and wheel to Ubon. Game for about anything, I volunteered, but was set back a little when I saw the thing tied onto the wing tank with rope.

I accepted the challenge and launched, but with a wheel hanging outside, I had to fly at 140 knots during the flight to Ubon. It was slow going. After all our diverts to Ubon, the air controllers en route could practically identify our pilots by voice transmissions. Assuming I was having some flight problems, mine asked why I was flying so slow. I told him he would, too, if he had a wheel tied to his wing tank. I explained the situation. We both had a good laugh, but by the time I landed at Ubon, the word was out. Some crazy Marine had a wheel tied onto his wing tank and was going to land at Ubon. During my approach, I noticed a large gathering of Air Force personnel standing out on the aircraft ramp watching my approach. After landing, they filtered back into their hangars and work spaces, probably disappointed they hadn't seen a crash. The mech that flew over with me pulled out his jack and tools, and in about thirty minutes, the bird was ready to fly. We tied the bad wheel onto the wing tank and went 140 knots back to Danang. The Morley Far East Transportation System had worked again.

Returning from a Navy strike one night, I was diverted to Korat, Thailand, for the first time and met a very interesting lady. We didn't hit Korat very often, but someone told us Verna's was the best place in town to eat. Verna met us at the door, and when we told her we were Marines, she wouldn't leave our side. She ordered a whopping, delicious dinner for us, and then sat down to talk about some beautiful Marines who had frequented her abode a couple of years previously. We picked her brain about Thai history, the King, her town, and just had a great evening. She was one sharp gal and a gracious hostess . . .

wouldn't let us pay for the meal either.

Liberty in Thailand was a lot of fun. We could wear our flight suits into town, the drinks were pretty good, and the food, when we ate, was passable. Most of our diverts were to Ubon. The Air Force guys flew out of the field, and from the looks of their security procedures, apparently were flying on some super-secret missions. Ubon also housed a squadron of Australian pilots. Our Major Tim Fallon flew his F-4 into Ubon one night and got tangled up with the boys from down under. Seems they were at the end of their tour, flying home by commercial air the next morning, and because they didn't have to fly their birds home, were bent on drinking all their excess booze before leaving. Tim and his R/O joined the party!

The R/O gave up about 0200 in the morning, somehow found his way out to the F-4, and went to sleep on the wing. Tim woke up at daybreak lying next to the officer's swimming pool . . . sans everything but his white T-shirt and boxer shorts. His party pals had absconded with everything else to take home as souvenirs. I believe his wallet, watch, and flight boots were stacked nicely by his head. He laced them on and headed for his bird. He woke up the R/O, poured him into the backseat, put on his own torso harness, and somehow flew back to Danang (probably blowing a 3.0 all the way).

After shut-down, Tim was too embarrassed to alight from the cockpit and asked a rather surprised plane captain to round up some clothes for him. He couldn't . . . but he spread the word . . . and then came back with a cameraman from our photo department. He got several nice shots of Tim climbing down from the cockpit. The rapidly gathering troops got quite a kick out of Tim's "underwear" walk to the hangar.

I finally got to Fam the F-4. Guess someone felt sorry for me, because I really wanted to fly the bird. The hop called for some low-level road recce photo just southwest of Danang. Man, this was fun. The F-4 was the real Cadillac of all airplanes to fly, and low-level flying was my piece of cake. In fact, while flying the RF-8 Crusader out of El Toro, California, I'd fly out to the desert area along the Colorado River above Yuma, spot a pickup truck making rooster tails in the dust down a dirt road, and make a few balls-to-the wall, ground-hugging passes at them. Stopped a few, too. While flying the F9F Cougar in the Advanced Training Command at Beeville, Texas, I flew out to a cowboy friend's rice ranch near El Campo and caught him out on his tractor, plowing a field across from his house. I came in behind him, then did a 180, and bore-sighted the nose of the tractor. He saw me coming, jumped off the tractor, and took off across the field running. I had a good laugh, called him after I landed back at Chase Field, and found out he hadn't even been home. I had buzzed his neighbor who had threatened to "kill the goddamn pilot of that jet, if I ever find him." Needless to say, I maintained a low profile when visiting my friend.

One night, while auguring around in Willy on a DMZ hop, I heard a controller come up on Guard asking any aircraft carrying ordnance to come up on his freq. There was an enemy chopper headed out toward Tiger Island, and they wanted to hit him on the return trip to NVN. (Tiger Island was a rocky, tree-covered island off the coast, about 15 miles east of the DMZ, controlled by the North Vietnamese.) I received permission from my controller to switch to the caller's freq, told him I had 20-mm guns aboard, and was at his service. He gleefully gave me a "wait one." Man, can't you just see the headlines: ol' Willy just shot down an enemy aircraft, the first one since it was the F3D, a night fighter back during Korea! Apparently, he had put in a request to use me, and when he came back on the air, gave me the rundown on my target. It seems this particular chopper had made a run to Tiger Island several nights in a row. It stayed down around 500 feet to avoid detection but could be heard and seen passing over some of our ships loitering between the coast and the island. There was a little moonlight, so I figured if I got down to 200 feet, I might be able to pick the chopper up on the horizon and shoot his ass down.

I was a little surprised, however, when the controller came back, wanting to know all the particulars about my bird, the guns, and my service. A rather disgruntled controller called back a little later, stating he couldn't use me, a Marine. This was an Air Force problem, and they would handle it the next night. Pissed? Yes . . . but thinking it over later, the higher-up who made that decision might have saved one Willy the Whale driver from flying into the ocean. NAH!

I made sure I was on station that next evening. I reported in to my friendly controller, letting him know I was available if the Air Force messed up. They did. I didn't hear another thing about the action that night, but the next morning, while reading through the early dispatches, my eye caught something about a Navy destroyer being hit by friendly rockets while patrolling west of Tiger Island. Seems one of the Air Force pilots had a radar lock onto what he thought was an enemy helicopter and fired his rockets . . . at the destroyer. One of the rockets was a dud, and the Navy became a little chagrined when they observed U.S. Air Force markings on the thing.

Big ol' jovial George Potter, our Grumman tech rep, and I were sitting out on the patio one night watching the movie. Pilots of a newly reported squadron were celebrating their good fortune in the club that night. All of a sudden, someone from Dog Patch lobbed a grenade over the fence. It landed on the opposite side of the club from where we were sitting and exploded in a great grenade roar. Man, you should have seen the new arrivees pour out the doors of the club (our side). The stampeding, uninitiated ones literally rolled through the patio area, knocking over tables and spilling drinks on the floor. One guy climbed right across our table knocking poor ol' George off his chair. As he slid back into his chair, he said, "Goddamn those wild animals. Don't they know, if you can hear a grenade go off, you're okay? It's the ones you don't hear that kill you!" The wild ones continued their dash some 30 yards from the club, stopped, and stood embarrassed when they saw all of the initiated old-timers sitting there laughing at them.

The highlight of my tour in VMCJ-1 was my assignment as the squadron S-3 (Operations Officer). Apparently, intelligence reports were indicating the NVN were stockpiling goods and moving troops into the northern part of the DMZ. They were expecting a major enemy attack through the DMZ to capture Dong Ha, Quang Tri, and possibly Hue. They needed full photo coverage from the DMZ north about twenty miles, and inland to the foothills, some fifteen miles.

Unexpectedly, we received a dispatch tasking us with the photo coverage with "block" times (clock times) to enter the area the next day. The hidden agenda seemed to be . . . you guys are the closest photo outfit to the DMZ . . . that's the only reason we are using you . . . if you need Air Force assistance or cannot obtain the desired coverage . . . please notify.

This was going to be a huge assignment, but the operation order specified the task would only last three days. It couldn't have come at a worse time. Photo-wise, we could cover the area with two F-4s and two medium-altitude cameras. Trouble was, we only had two cameras, and they were both down, awaiting parts from the states. The alternative was eight planes flying two flight lines apiece with low-altitude cameras, putting our pilots at risk. (None flinched.) During a conference on the subject, we decided it would require a max effort by all hands, but coverage could be accomplished if the aircraft and cameras performed well. (Our photo department and maintenance crews wouldn't be getting much sleep.)

After laying out the plan for our CO and then the Wing Chief of Staff, we went back to Saigon with a "gulp" roger. We didn't dare tell them how we were going to accomplish the missions. The assigned block times allowed us to enter the area with two aircraft for one-half hour, one in the morning, one around noon, then one in the afternoon. We planned on sending all eight planes at the same time, hopefully during the morning block time. This would allow our people to start work on photo and maintenance immediately after the birds returned.

The first day, we obtained full coverage of the area. Same the next day. The weather turned bad on the third day. We had to shoot it in the afternoon, but still got full coverage. By now there were a lot of fannies draggin' around the squadron area, but the ordeal was over. Not so . . .

Guess the big shots liked our work, because that night we received another Op Order requesting another day's coverage. Someone up the line decided there were so many targets in the area, we had to start a massive air attack on the DMZ. Everything from artillery to B-52 strikes were scheduled, and they wanted our photos for bomb damage assessment (BDA). We gave them 100% coverage. Luckily, one of our medium-altitude cameras came online, so our flights dropped to four planes, but the photo guys still had to process just as much film. Tasked for a fifth day—another 100%. The requests didn't stop, but our other camera came "up" and greatly improved our situation. On the eighth day, it all came to an end.

Every person in the squadron contributed to the venture, many with hardly any sleep. We met other squadron commitments, including photo, had obtained full coverage of the DMZ shoot each day, and had accomplished a task probably never equaled again by a VMCJ Squadron. Yea Team! Honk Honk!. . . and all that good stuff, and yet, I don't remember receiving one accolade for our work . . . not even a thanks. We didn't need it . . . Hell, we were VMCJ!

By September, I had been in country long enough to rate an R&R (rest and recreation). It didn't take me long to get all the paperwork out of the way and soon found myself on an airliner headed for Australia. Just to show you what a great, laid-back country Aussie land was, we landed first in Darwin (up north). After sitting in the airport for an hour and a half, I queried one of the pilots on the delay. The answer was, "They don't open the Sydney Airport until eight o'clock in the morning. We've got another thirty minutes to wait before takeoff." We arrived over Sydney about fifteen agonizing minutes early and had to circle until they opened the field. I hit the ground running.

The first thing I noticed as I entered the hotel were those long-legged, beautiful Australian gals walking around in miniskirts. A guy sure didn't have to expend much energy getting a lovely lady out on a date, and what beautiful ladies! (That's all I'm going to say about that subject.)

After using those smelly, outside toilets for the last six months, just flushing a toilet was pure joy!

The Australian folks really treated U.S. servicemen great. I guess it was a throwback from WWII when the island was flooded with Marines, because they took special note of our Marines. Each hotel housing R&R servicemen fortified the lobby with three or four tables. Manned by local women throughout the day, servicemen could talk to them, specifying what they would like to do while in Sydney, and the gals would make the arrangements. On the way back to Vietnam, I talked to several about their experiences. One had even taken a guided hunting trip into the outback hunting kangaroos. Most of the kids, however, knew the key words . . . I'd like to meet and visit an Australian family. It seems most of these families had good-looking, datable daughters, fed the troops outlandishly, got them drunk as skunks, and wouldn't take a penny for their trouble.

I owned some racing quarter horses at the time, so I was very interested in visiting a stud farm and a race track. When I expressed my desires, the lady got a funny look on her face and said, "We've never had a request like that." She soon found another lady who knew someone who had a horse. She wanted time to work on my problem, so I went into the restaurant to get some breakfast. As I was finishing, the lady came running up to my table, gave me a slip of paper, waited while I paid the bill, took me out of the hotel, hailed a cab, shoved me inside, and said, "give that paper to the driver. I found someone who had a horse who knew someone who had a horse who knew someone . . . and I think you will be happy with the people at that address. I don't even know them, but I'm sure you will be all right."

The cabbie eventually pulled up in front of a large office building. I went inside and found myself standing in the office of one of the most prestigious horse auction agencies in the world. No way were they going to have anything to do with a little operator like me. Wrong! They were the most gracious folks in the world. Within minutes they had me lined up to go out to the racetrack the next morning to observe morning workouts (at 0500), and a visit to a thoroughbred stud farm was arranged in the afternoon. On top of that, the owners invited me to go with them out to the Sydney racetrack where races were starting in about an hour. At the track, I was introduced to a number of people who, once they found out I was a Marine jet pilot visiting from Vietnam, invited me out to their homes for a visit. (Had to turn them all down—too many lovely ladies running around town.)

As the horses came out on the track for the first race, I noticed they were headed in the wrong direction. I had forgotten that, down under, they run the wrong way (clockwise) around the track. Much to my delight, as I went to the paramutual window to place my first bet, there were no lines forming. Before placing my bet, the lady said, "Wouldn't you rather place your bet with a bookie? The odds are better. They are right over there behind the trees." Sure enough, I had only seen the bookie situation in old English movies, and here it was, spread out right in front of me. I had a fun day, and after the last race, my new friends dropped me off back at the hotel.

The next day found me in the dark at the race track. One of the most famous trainers in Australia was waiting for me. As we watched the horses working out on the track, we got into some real interesting discussions about race horses. I learned a lot from that visit and made some wonderful friends.

That afternoon I hopped a train, traveled out into the country, and was met at the station by a beautiful lady. Her husband (shucks) owned Webber's Bread Company and ran a big stud farm. What a beautiful place. It would compare to some of the best farms in Kentucky and California. As we entered their yard, a dog and a little kangaroo, about a foot tall, came bounding up to meet us. I was amazed. He was just a house pet. You could scratch and pet him just like a puppy. The owners told me the little guy was full grown, unlike the large beasts seen in the movies.

Along with the horses, these folks owned a pig farm. They took me into the large birthing barn, the cleanest pig place I had ever seen. Everything was automatic. The heat came on when required, the lights when it got dark, ground sprinklers came on at certain times to wash away the manure into little trenches where conveyer belts transported the waste out of the barn into a waiting truck—magnificent—and there were hundreds of pallets filled with excess or day-old Webber's bread to be fed to the pigs. Some kinda operation!

Their horse operation was certainly first class. As we rode out in his station wagon to look at his pastured horses, I volunteered to ride in the front seat and be the gate opener. Rain started coming down as we went from pasture to pasture, and by the time we arrived back at the house, my dress shoes were covered with mud. Rather than track mud into the house, I took the shoes off and placed them on the backdoor step. We had a good visit, but when it came time to leave, I went out back to retrieve my shoes. All the laces had been removed. The lady of the house laughingly stated, "I told you not to leave your shoes outside. Our kangaroo has a bad habit. He eats shoelaces, right out of the shoes!" We all had a good laugh over the incident but when I arrived back at the hotel, still with mud caked on my shoes (tough mud), I walked into a lobby filled with folks in evening gowns and tuxedos. I was just a little embarrassed waiting at the elevator. There were a lot of "Oh, I feel so sorry for that poor man" looks from the ladies and quite a few snickers from the gents.

During my week in Sydney, I met a wonderful lady—a second-grade school teacher. I guess she shared several stories with her pupils about our conversations, because it wasn't long before I started receiving letters. Here are some samples:

"Dear Major Draay I wish you come back from Vietnam. When I grow up I am going to be a pilot like you. I will like to fly in a jet like you. from Michael"

(I sent him a pair of aviator wings); then,

"Dear Major I thank you for giving me the army badge I will be proud of it and clean it every day. I hope I will not lose it. I hope you will have a Happy Christmas with your friends at Vietnam Your loving friend Michael."

Then a Christmas card:

"Dear Major Draayer I thank you for giving me the badge. I'm sending you a card for Christmas and for all your friends and I hope that you kill the Tager and I hope that none of you will not get shot. You are one of the best men I ever see." (Tager was from the tiger story, printed in the *Stars and Stripes.*)

It didn't take long for:

"Dear Major Draayer I hope you are well and unhurt. How is the F 8 Crusader going. I hope you have a very nice Christmas and wish all the other men a very nice Christmas too. The wings that you gave Mc Michael are very nice. I'm sorry that one of the marines got dragged away by the tiger Love STEVE"

"Dear Major Draayer I hope you bring me one of those Army badges. And I also hope you have a Merry xmas. I hope you Don't get shot please get me one of those badges. Love Raphael"

After these hints about sending them wings, I rummaged

through my possibles, found a 1,000-mile-an-hour pin, several Crusader pictures, badges, etc., then bought some more samplings from the PX, and sent them off.

"Dear Major I hope you don't get shot by the viet cong All the children saw Michaels badge It is very nice I hope I get an army game for christmas I will call some viet cong and, others united states Marines If I was in the army I would shoot every one down dead We play Marines at school and I wish you a Happy christmas you loving friend Rodney."

"Dear Major Hope you are well. I'm sorry to hear that one marine got draged away. I like the Crusader wings. Hope you have a happy Christmas. I came first in the grade. Love Richard"

"Dear Major Draayer I hope you win the war I thank you for that pin We are playing air force. Michael is Major Draayer. your loving friend Murray"

"Dear Major Draayer I hope you on fighting well in vietnam and if you get Shot when I grow up I mite be able to be in the army air force. I hope your very well in vietnam From Nino"

"Too Dear Major Draayer. I hope you are well Major Draayer I'll Dive in a Jet to shoot the noth Vietamese down with my machine Gun and blow them up with a hand grende when I Grow up and am won of the Marines. Love from Adamadios"

"Dear Major Draayer I hope you are having a nice time in Vietnam. If you get shot when I get in the army I will shoot evry vietcong in the head. Do you feel well From Murray would you send us some more photographs please. Good luck (turn over) Major will you send me a piece of your uniform please The End"

Most of the letters had U.S. flags, Crusaders, and Marine symbols drawn on them. A great bunch of kids. The Aussies are great people . . . the best!

Arrival back at Danang found me toddling off to Wing G-1 to inquire about the prospects of becoming an O-1 driver. He said, "We've got openings for volunteers right now and don't have one single applicant." I told him to sign me up, called my VMCJ-1 Commanding Officer for his okay, and six days later I was on my way to Quang Tri.

Chapter 6

Quang Tri

I caught a KC-130 from Danang to Quang Tri on September 25. My fellow passengers were a bunch of grunts. They had been in action down around Chu Lai for about three weeks and were now headed for Dong Ha to regroup and go back into the bush. This was a wild-looking, tired bunch, with that blank, 1000-yard stare in their eyes, and my guess was that they had seen a lot of tough action. Now, I'm a firm believer that Marine Corps Air serves only one purpose: support that grunt in the field. On that plane ride, I renewed my vow to do whatever it took to help these great guys and their fellow Marines. That's what it's all about.

When I finally found MAG-39's Admin Office, I turned in my orders, not knowing what my new assignment might be. I was told the O-1 Bird Dogs were co-located with a Huey helicopter squadron . . . VMO-6. By the looks on their faces at the Group, I don't believe they were expecting a senior Major to show up as a replacement Bird Dog pilot. While they were sorting the seniority thing out, I wandered over to VMO-6's Operations Office (OPS). The first guy I ran into that I knew was none other than Captain Mike Whitehouse, an old compadre from my grunt tour at Camp Pendleton. He introduced me to Major Ken Buske, another great guy. One of them stuck an O-1 Operational Handbook in my paw, and I was on my way to becoming a Forward Air Controller (FAC), and I got checked out by the best in the business. Mike and Buske had been a force at Khe Sanh, where statistics tell us 95% of the air strikes were controlled by FACs, not to mention the use of artillery by them and their AO counterpart in the back seat. The FAC pilots of VMO-6 had all been in the action at Khe Sanh, the many Operations in I Corps, and performed brilliantly. I enjoyed many of the "daring do" tales that were told nightly at the club.

Once the seniority list was figured out, I was begrudgingly assigned as the Executive Officer of VMO-6. Hell, I wouldn't have cared if they just put me down as Squadron Pilot! I just wanted to get into . . . my kind of war. The Commanding Officer of VMO-6 soon became my most unforgettable character. Major Fritz Zander (just a few points senior to me in rank) had spent considerable time flying the Presidential helicopter during tours of duty in Washington, D.C., and, unlike most of us, had even flown in and out of Hanoi, the latter being when he flew from one of our carriers in the Tonkin Gulf into Hanoi to heli-lift a defeated French Army out of Vietnam. He was bald as a bat and could keep you literally doubled up with laughter with his rendition of a WWII German U-boat commander or Stuka pilot. Fritz usually wore a monocle over his eye while sitting at the bar in the O'Club. If an officer became unruly, Fritz would walk up to him and shout for all to hear, "Take this man to the showers, but first tattoo his arm!" It always got a big laugh and usually subdued the offender.

The Marine Corps designation for an airborne FAC was TAC(A) (Tactical Air Controller Airborne). We also had Forward Air Controllers (Marine aviators located with our ground forces who controlled air out of their radio mounted jeeps or hand-held communications gear); however, we liked the name FAC, which was generally used by all the services to designate the pilot flying the O-1. The guys in the backseat had various nicknames. They were officially labeled AOs (Air Observers). My favorite (respectfully) was "backseater." This did, however, create a little confusion with our brothers on the ground. Many thought the AO controlled air, arty, etc. The FAC (pilot) always controlled air strikes and most of the other missions. Our backseaters were Marine 1st Lieutenants or Warrant Officers, mostly with arty backgrounds or some directly from a recon outfit. (They had to have the biggest balls in the world to volunteer as a backseater in the O-1.)

The O-1 Bird Dog was a real tiger to fly—80 knots in a climb, a glide, straight and level, or flat on your back. I say that, but for me it was a joy to fly. Almost my entire flight time had bent spent in jet aircraft. The Bird Dog didn't need a starting unit, a gear handle, or a 100-item checklist to go over prior to takeoff. Hit a switch, crank her up, and you were ready to go free. You could go anywhere and do anything your heart desired. We very seldom had detailed missions. You had rocket tubes installed under each wing, and the gun sight was the third rivet up from the bottom on the metal support between the split windscreens (windows). You had to lean way over to the right to get it lined up, so to be a little more accurate, I placed a black grease pencil dot on the windscreen right in front of me. After each rocket fired, I would move the dot to another location, in essence, bracketing the rounds until I had a respectable gun sight. Only problem was, other pilots started doing the same thing, and soon there were dots marked all over the window, eventually with initials next to them signifying ownership. Then some unbeknownst plane captain would clean the windshield, and all would be lost.

Being a top winger, the Bird Dog provided excellent visibility—it even had windows on the top, right over the pilot's head, so one could see the ground in a banked turn. For good listening, the back seat windows were removed. This was a little breezy on the guy back there but, man, when someone fired at you, he was going to be heard and quickly spotted. The rudder trim was a little piece of metal attached to the back of the rudder. When you flew a bird that required rudder during the hop, after landing, you hand-bent the piece of metal to where you thought it might suffice for the next hop. The only problem was we couldn't seem to determine which way to bend the piece of metal, left or right, so we flew with a heavy foot on the rudder a few times. You could land the little bugger on a dirt road, or the tops of trees, if necessary. All FACs felt if they got hit and had to go down, if a suitable landing spot was unavailable . . . look for some, soft, tall trees to land in. We felt we could mush the thing into the tree tops at about 50 knots and walk away. Don't laugh, it was done a couple of times, and there were a lot of trees in Vietnam. Now you can laugh.

Literature on the O-1 states:

"It has a 213 hp Continental air-cooled engine, capable of 100 knots, loitered at 80-90 knots and was unquestionably one of the most important and hard-working aircraft throughout the Vietnam campaign. Features included all metal construction with control surfaces stiffened by corrugated skins, a high wing braced by single struts, with slotted flaps, spring cantilever main landing gears, and a well-equipped cabin affording an almost perfect all-round view. Production continued until 1962. In the vital FAC role, in which the O-1 served the US Army, Air Force, Marines, VNAF and Royal Lao AF" *(doesn't mention the birds flown by Air America).*

"The aircraft might be flown solo or with a second crewman. Targets would be studied and assessed from very close range, often in a manner deliberately intended to draw hostile fire. Targets would then be marked by smoke rockets." *(We used Willy Peter—White Phosphorus—rockets.)*

"The main drawback of the O-1 was its lack of armor, and self-sealing tanks." (Aircraft of the Vietnam War. Gunston.)

Most of the pilots who flew the O-1 dearly loved the little bird and the mission it performed. A few wore parachutes, but most used them to sit on. (We never flew high enough to use them anyway!) We also had large, thick, steel plates designed to strap on and fit over the chest. Most stuck them under the parachute for a little butt protection.

Another publication wrote:

"…in awe of the fleet of helicopters and jets, equipped with a terrifying arsenal of weapons at our beck and call."

We were an excellent source of raw intelligence, could take pictures with hand-held cameras, and, in essence, operated in our own potent little world. I'm not sure of the origin, but one publication writing about the FAC mission pictured:

"The FAC makes radio contact with the oncoming attack planes and talks them in over his head. When they have him in sight, the FAC rolls in on the target and fires a smoke rocket. If the rocket missed, the FAC tells the attacking planes how far from his mark and in what direction to unload. After the smoke rocket, the napalm won't be far behind. The VC hate the FAC. It is he who fingers the death bringers. If they catch him on the ground, they kill him. Near Pleiku, one FAC was literally skinned alive by the VC. More FACs have been lost in Vietnam than any other type of pilot."

(I weighed in at 210 in those days and would have made a huge "bear skin rug.")

Which brings me to Mike Whitehouse's theory on why O-1s seldom got hit by ground fire. It goes like this: Any light aircraft like the O-1 is affected by the wind. To fly a straight line from A to B requires a crab into the wind to stay on that straight course, thus the O-1's nose is continually pointed away from its track. Gunners on the ground know they could not hit the little bird by aiming directly at it (it flies too fast), so, they aim in front of the nose where, in actuality, they are going to miss because the path of the O-1 is slightly off the nose. The O-1 drivers firmly believe this and feel they are invulnerable to ground fire—and, if you believe this, go out to any local airport, park near the landing end of the runway, and see if you don't think you could hit one of the light aircraft, any aircraft, with a rock! The theory has a few holes in it, but it was a nice crutch to lean on.

In general, the enemy seldom fired at an O-1 while we cruised overhead. They knew if they fired and you spotted them, in a very short time, all hell would be raining down around their ears. The little guys held off until you rolled in to mark your target, and then all hell broke loose.

The procedure for obtaining air over a selected target involved a lot of "paperwork." Once you had a good target, you had to locate it on your map by coordinates, call it into a ground controller, requesting specific ordnance desired, etc. He, in turn, checked with other agencies to see if there were any friendly troops near the target and sometimes determined if the target was worth hitting. If the requested ordnance and birds were available and the request was approved, the assigned birds would come up on your frequency and report with all their particulars (ordnance aboard, time available on target, etc.). You then gave him the required target information, your position, the run-in heading, how you wanted the ordnance dropped, and any number of things to make their runs successful.

Most of the Marine A-4 (Skyhawk) drivers knew the general I Corps area landmarks by heart. Give them a "meet me ten miles northeast of the Rockpile," or "three miles south of Khe Sanh," or ". . . on the DMZ just north of Con Thien," and they would be over your position in no time at all. The uninitiated were given long or short ADF radio steers until they were overhead. I found a great reluctance on the part of jet pilots to let down from their fuel-saving 20,000- to 25,000-foot altitudes. We FACs had difficulty finding them at that altitude, so we usually dropped them down to 10,000 feet where they were readily spotted. You could then direct them to your target simply by voice commands (for instance, if he was headed away from you, "give me a 180, drop down to 3,000 feet . . . stop your turn, I'm right on your nose . . . come a little left . . . stop your turn . . . I'm at your eleven o'clock, low . . . I'm in a Bird Dog with white paint on the tops of the wings…I'll rock my wings until you spot me"). Very seldom did the birds have to make more than one pass.

When the attack aircraft had you in sight, you rolled in, marked the target with Willie Pete, performed a quick, 360-degree turn to parallel his run in and, if Lead (flight leader) was on heading and you were satisfied he had the target in sight, you cleared him to drop. The same procedure was used with Two (wingman). From that point, it was simply giving them the proper corrections from the last run. During a three-hour hop, if you ran out of Willie Pete rockets, you threw smoke grenades out the window. Sighting was simple: dive in and at twenty or thirty feet above the ground, as your left wheel was about to pass over the target, you dropped the grenade. Needless to say, there were a few overs and unders, the only dangers being taking ground fire or having the guy in the back seat throw out the grenade. New AOs would sometimes hold the grenade out into the slipstream, palm forward, release the spoon, and open their hand, only to find the grenade still pressed firmly in their hand by the slipstream. The screams blazing from the pilot's throat usually jarred them enough to get them to twist the palm downward, letting the grenade fall free.

Arty requests were handled in the same manner as requests for air, except the "paperwork" was less time-consuming. The AOs frequently handled the air/arty radio requests, allowing the pilot the freedom to keep the target in sight. Once you received target clearance, you simply had one gun fire a Willie Pete arty round, observed where the round hit, had the pilot or AO make corrections until you had a target bracketed, and then called "fire for effect," when several guns fired.

I soon found a bed in a Quonset hut that was called the Majors' Hut, probably because there were eight Majors living in the place. The hut was a great place to live. (We even had television, but it seems all we received were the local news and the *Glen Campbell Show*.)

All the Majors got along well together and, best of all, we received an abundance of goodies from our families back in the states. Man, it was just like Christmas when we received packages from home. We usually waited for most of the Majors to arrive at the end of the flying day and then would sit around opening the packages. We shared everything. Especially anticipated were the packages for one of the guys whose father ran a deli back in New York. (Those kosher salamis were great.)

My sister in Denver, Colorado, sent us a lot of chocolate chip cookies that were the best we ever tasted. Every time I received a package, everyone asked me if it was from my sister. They usually came in big boxes, two hundred or so cookies to the box. One day I received a smaller box, took it into my office, and placed it over in the corner. I was determined to save the box to share with my roomies later, but my conscience wavered a little. Not being inclined to stand the temptation any longer, I opened the box, which contained probably sixty. I took a couple in to Fritz, had four or five myself, then closed the box. Not for long. I absolutely loved those cookies. After Fritz left his office, I tore into the box and ate every damn cookie in the thing! Later, in front of all my roomies, Fritz, bless his little pea-pickin' heart, asked when I was going to roll out my box of cookies. I could have killed him. They almost killed me. I took a lot of heat for being the greedy one.

The "shitter" (as we called them) was a real comedown. I had de-graduated from a nice 10-holer at Danang to a 3-holer, and the thing was located within a stone's throw of the Quang Tri River. Due to the width of the river, at least there would be no thrown grenades from our Dog Patch, but it was sure within rifle range. At least we sat with our backs to the river and wouldn't see the round coming . . . Lord, what a thought: killed in action while reposing on the middle hole of a 3-holer!

The day after I arrived, Mike gave me a Fam (Familiarization) hop, then an aircraft check-out hop, and pronounced me qualified in the Bird Dog. Lord, it was fun to fly. Two things were instilled into my brain on my Area Fam with Buske. As we flew over Khe Sanh, he said, "Never get caught below 1,500 feet over Khe Sanh. You'll get your butt shot down by ground fire." We had been flying about an hour when he suddenly asked, "Where are we? Give me the grid coordinates for the area we are over right now." Hell, I hadn't been following that closely on my map. I wanted to impress, so picked out a couple of streams, matched them up on my map and gave him the coordinates. He just laughed, "You're at least a mile off. You've got to know exactly where you are at all times!" I felt like an idiot. The first few days, when I didn't make the flight schedule, I would jump in the back seat and fly with Ken or Mike to suck every bit of experience they chose to divulge.

After the Khe Sanh siege, little enemy activity was observed in the I Corps area. Our O-1 pilots who flew Khe Sanh hops seldom went below 1,500 feet due to enemy ground fire. The standard question amongst O-1 and Huey drivers at debriefings was, "How many did you see today?" After Khe Sanh, it soon became, "Did you see any today?" By the time I arrived in the area, a pilot could fly at tree top level over Khe Sanh and, in general, our whole operating area. Before Draayer, our pilots could readily see troops in the open, un-camouflaged fighting holes, etc. During my time, you saw very little. There just wasn't any traffic. We had won the war of the DMZ! So we had to revert to our experience.

After flying repeatedly over an area, you could pick out signs of enemy activity. Over a period of a few days, you started to pick up little nuisances that said "enemy." We would then call for arty to interdict and blow a few holes in the canopy, hoping to see signs of the enemy. They had been regrouping for some time now, particularly up north where absurd bombing halts had been restrictive. Their main supply activity hadn't been directed through the Marine Corps DMZ area but had been directed in an end-around move down the Ho Chi Minh Trail to our west—in Laos.

Ken Buske was a real character, had been wounded, I think by being hit in the toe by shrapnel on his way to our bunker outside the Majors' Hut during a rocket attack. He kept an M-16 rifle, flack jacket, steel helmet, flashlight, and .38 Special hung up near his bunk. As soon as the first enemy rocket hit or a siren blared, you better not be in his way. He somehow got all the hardware flung around his body and even if he was half awake, he'd be the first one out the door and into the bunker, usually in his slippers and shorts. He loved the O-1 mission and was one damn fine professional.

On one hop with Buske, I spotted two gooks running, then crouching, up a road near one of our fire bases just below the DMZ. I pointed them out to Ken who immediately called for air—none available. Called for arty—none available. He was mumbling misgivings about our support when I called and said, "Hell, I've got my M-16. Let me try it."

I had heard many derogatory remarks about the M-16 rifle. The Army was experiencing a series of the weapons jamming. A few Marines experienced the same difficulty. Before my introduction to Vietnam, I had read a study put out by the Corps stating the primary cause of the jams was an unclean rifle. A Marine keeps his weapon in top condition, so the problem was of a minor nature in the Corps. I had become a champion of the weapon and had argued the point at the Naval War College with other officers, namely our Army friends. *(Post-Vietnam publications concerning the Corps' participation in the conflict indicate Marines experienced the same difficulties.)*

Both the pilot and AO carried M-16s, Velcro-taped in the corner of the cockpit. We had already received clearance to hit our target, so Ken said, "Go ahead, but there's only one problem: you can't shoot the thing like a normal rifle. If you do, the shells eject, hit the pilot on the head, or drop between his back and the seat, and they burn like hell." Firing out the right side of the aircraft was no problem. When shooting out the left side of the plane, the spent shells ejected up and to the right. It was more comfortable to shoot it left-handed, though, and a little more accurate; however, I could see where the pilot could get a little ticked with shells raining down on his neck and backside. I was told to turn the rifle on its right side (so the shells would eject down onto the floor of the aircraft), hold the barrel with my left hand against the bottom of the window sill, sight, and fire. Picture this: you couldn't hold the rifle up to a shoulder; you had to hold it against your chest, with your left hand out on the barrel guard, and then place your right arm up and over the top of the weapon to reach the trigger, all the while trying to get a sight picture—a real acrobatic maneuver performed with great difficulty.

By the time I vaguely mastered the thing, I fired one round and the damn weapon jammed. I took out my survival knife, pricked out the jammed shell, tried again, and after another round, went hurling toward the gooks. Lord, it jammed again! My respect for the weapon rapidly diminished. I threw the damn thing on the floor and told Ken to give me his rifle. His worked! Through a combination of "buck fever" and wrestling with the firing position, I had lost the gooks. Not Ken—he said one had been hit and was laying in the road, the other was hiding behind a large tree, and he told me to . . . simply hit the guy in the head when he peeked around the tree. I couldn't see the gook lying on the road . . . probably looked right through him. I considered Ken crazy as hell if he expected me to even hit the tree, let alone the guy peeking around it. I guess Ken had the same thought and told me to just fire at the base of the tree—maybe we would luck out and get a kill. My first shots missed the tree by a large margin. I had forgotten to take into account the forward movement of the aircraft, or lead. Then things got a little wild . . .

"Draayer, don't you have sense enough to throw in lead for our forward motion? God, where did you learn to fire a rifle?"

"Not out of a damn airplane! Ken, if you'd keep this damn plane steady . . . how can you expect me to hit anything with you bouncing all over the place?"

"Look, I'll come around the tree again and give you another shot. Hit something this time, at least the tree. You might get lucky, hit a limb, and it might fall on his head and kill him."

"Well, we are here, and I still don't see him. Have you got him in sight?"

"I've got him, but he's hiding on the other side of the tree. Just a minute, I'll fly around to the other side and you can pick him up."

"Roger."

After several circles around the tree, the only tree in the area . . .

"Draayer, when I circle, he just keeps circling on the other side of the tree so we can't get a shot at him."

"Ken, you dummy, fly right over the top of the tree, bank over, and I'll get a shot!"

We performed the maneuver. I didn't see the enemy, but fired several rounds at the base of the tree, rather accurately, I must say. We went back for another pass, but Ken said he had seen the guy limp off into the nearby bushes. He couldn't spot the one lying on the road but figured we got both of them. I seriously doubted the latter. Ol' Ken was a great gentleman and fun to be around.

On 30 September I was in the backseat with Buske and hadn't controlled arty yet, so he decided to give me a little instruction. Ken picked a target way out on the Laotian border below Co Roc Mountain. Right after the marking round, someone popped a red flare on the ground. This meant there were friendlies in the area. Ken figured it was a recon team, came up on their frequency, profusely apologized, and we canceled the mission. (The 9th Marines requested an arty mission up on the DMZ later during the hop, so I received some live instruction.) Now, I'm not knocking Ken for dropping on friendlies. God knows, we were miles from any of our regular troops, and who would believe an isolated area like this would be inhabited by a recon team? I had much more to learn about the recon folks. Dropping on friendly troops did happen but very seldom by our FACs. We worked close to our troops as a matter of course, and the close contact experience really paid off. The system wasn't perfect.

When I first arrived at Quang Tri, we had more than our share of mice running around our kitchen. I sent home for six mouse traps and two large rat traps. (Some of the larger rats would have eaten the little mouse traps for breakfast.) When they arrived, I set them all over the kitchen floor and in the cupboards. The mouse war had started!

We caught six the first day. Someone hung a sheet of tablet paper on the wall entitled MOUSE KILL: Killed In Action. We killed 27 mice within two weeks. The scorekeeper kept our running score of KIAs, specifying each as:

3 mice. . . . .DBT . . . . .Death by Trap

1 mouse . . .KBB. . . . .Killed by Broom

1 mouse . . .KBRT. . . .Killed by Rat Trap

1 rat . . . . . .KBMT . . .Killed by Mouse Trap

1 rat . . . . . .KWS . . . .Killed with Shovel

The coup de grâce . . . one night, as we all sat around socializing after dinner, one little teeny weeny mouse ran out of the cabinet and stopped right in the middle of our circle. He seemed to become very disoriented and flustered by all those "White Eyes" glaring down at him. The poor little bugger didn't stand a chance. Several clodhoppers slammed out at him, crushing the poor little fellow to death. While he lay there, we debated how he should be counted on our KIA board. It looked like STD. Stomped to Death was going to win out, but as we looked down at that wretched little crushed body lying before us . . . someone suggested:

1 baby mouse . . . CS . . .Committed Suicide!

On 6 October, one of our guys bought the farm. He was new to the O-1 program but was an experienced A-4 pilot. Some of our troops were taking fire from gooks in the mountains just south of the DMZ near Helicopter Valley. Our guy was assisting with air, when he suddenly dove into the ground. The ground troops quickly moved to the crash site but found both bodies badly burned. We flew a couple of our O-1 mechanics out to the crash site. They later told me the bird had gone almost straight down and, except for a few parts, nothing was salvageable. They also observed a couple of bullet holes in one of the wings, so we knew the loss was from enemy fire. The guy in the back seat was a favorite backseater for me. He was a recon officer with a great attitude and a lot of ground "in the bush" experience. I loved to fly with him, feeling if we ever went down together, I'd just grab his shirttail and tell him to find some way to get us back home. I expressed my feelings in a letter to his folks. We really missed him.

Later in the month, we lost another Bird Dogger. A 1st Lieutenant launched on a morning hop in CAVU (clear air visibility unlimited) weather, and we never saw him or his backseat again. About a month later, someone sighted some wreckage in Laos believed to be an O-1. We flew some folks out to the crash site. They determined it was an O-1, but not one of ours (Army).

There were days when the weather was so bad, we simply couldn't get out into our operating area. Lord knows we always tried, because the gooks really liked to travel without us looking down at them. Coastal areas were not much sweat. You could fly under a 500-foot overcast without worrying about getting back to the field. Our west area was another matter. The normal routes to get up into the Khe Sanh area in bad weather were south, up the Quang Tri River, through the Ba Long valley to Vandy Base, or north to Dong Ha, up Highway 9 to Vandy. Just west of Vandy, the mountains rose substantially, leaving only one way to proceed: right up Highway 9. The canyon narrowed in places and with a low overcast, you frequently had to fly right down on the road, sometimes right down on the river bed.

No sweat going out. If the valley was socked in, you turned around and went somewhere else. However, if you managed to get under the crud in the narrow valley and out to the Plateau, getting back home was something again. Now, the Bird Dog is not known for its instrumentation capabilities in bad weather, and it had absolutely no navigational aids. If the Highway 9 valley was socked in, and I mean fog right to the ground, you had to cross your fingers, climb up through the mess, run a time distance check on a heading back to Quang Tri, and hope you could let down through the overcast somewhere around the Base.

It happened to me twice. The first time, I broke out of the

overcast at roughly 1200 feet about two miles west of the field. The second time, the tower reported the overcast was down to 300 feet, visibility better than three miles. I didn't want to let down over the field where other aircraft might be flying, so I timed my flight to arrive over the Gulf of Tonkin and let down through the fog, hoping to be over the water to avoid any small hills, towers, etc. I broke out at about 400 feet and made my way VFR to the field. I'm not sure how the other guys did it, but I know there were a lot of flying rules broken just to accomplish our missions.

The monsoon season was upon us in October with driving rain for a solid week. The Quang Tri River rose out of its banks, right up to our base camp, and made quite a swimming pool out of our bunker. Now, I've mentioned Buske's prowess at donning his fighting gear and being first out the door to the safety of the bunker. This night, the base took some enemy rocket fire about midnight. He was blissfully sleeping, but when that first rocket hit, he arose out of his stupor, climbed into his fighting gear, and headed for the door. Fritz and I were climbing out of our beds as Ken went flying by. (Rank has its privileges. Our beds were the closest to the door.) As Ken exited the Quonset, I looked over at Fritz and said, "Stand by for the splash!" Fritz said, "We better go fish him out—he'll drown with all that gear on." We took a quick look to see if Ken had sunk to the bottom. Luckily, while going down the steps into the bunker, Ken hit the cold water, which revived him enough to stop his plunge into the four-foot-deep water. (He would have been listed KIP—Killed in Pool.)

On one of my first solo hops (no backseater), I was taking in the sights up on the Ben Hai when some of our troops on the south side of the river called, saying they were taking fire from north of the DMZ. They thought it was mortar or possibly artillery fire. I quickly flew up two or three miles north of their position but was chased out by what sounded like a machine gun. I climbed to about 1,000 feet, went back, and got chased out again. I finally climbed to 1,500 feet and was able to loiter over the area without taking fire. Each time I flew over the area, the rounds falling on our troops ceased. (Many times, just our presence over the enemy would make them stop firing. They didn't want to be spotted because they would soon have snake and nape (snake eye bombs and napalm canisters) raining down around their ears.)

We were not permitted to fly north of the Ben Hai unless fired upon, so we didn't normally carry maps of the area. Although a little inexperienced at picking targets, I found a few rather level areas with the brush cut back, possibly launch sites for mortars, so I called for air. An A-6 was soon on station, and as we were doing the "paper work," the pilot asked if I wanted a spread pattern. I didn't have the slightest idea what a spread pattern was. He told me it meant simply spreading the bombs along his flight path instead of dropping them all in the same area. Sounded good to me, and his run in was parallel to our troops, so I finally cleared him to drop on target. Lord, was I surprised when his bombs dropped about four hundred feet apart. I put a stop to it, called him back, and although I didn't have any coordinates, I talked him onto the target, marked it, and he thoroughly pasted the area. I checked with the ground commander during the next couple of hours and they hadn't received any more fire, so I thought somehow we had lucked out and hit the right spot.

While sitting at my desk the next morning, Fritz regaled me with his experiences taking the French out of Hanoi and flying presidents in and out of the White House landing pad in Marine One. I was thoroughly intrigued and thought it would be great to fly with Fritz just to say I had flown with the guy who had "flown Presidents." Fritz suggested it was time for me to become acquainted with the helicopter side of the Squadron. He had a hop scheduled, so I jumped in and played chopper copilot. We got shot out of the DMZ. Thanks Fritz. (Incidentally, it was over the same troop area that had taken fire the day before, so I guess my bombing wasn't too successful.)

After returning from this hop, I found out Capt. C.D. Saunders, a fellow Bird Dogger (a damn good one—looked like a Mafia assassin to me), had run out of gas up near Khe Sanh and had mushed her into a field of eight- to nine-foot-tall buffalo grass, causing the little bird to flip up on her nose. Sanders and his back seat were plucked out of the area by chopper within minutes, and a reaction force was heli-lifted in to guard her. Some of our mechanics flew in with them, took the wings off, tipped her right side up, loaded her into the back of a chopper, and brought her back home. The bird was back flying a few days later with not much damage (O-1s were tough and efficient). The Bird Dog wasn't very kind sometimes, though, as it didn't have fuel gauges!

Saunders had only been out two hours. Most of our hops were flown from two and a half to three and a half hours, and most were of the three-hour variety. We also had an O-1 run out of gas while taxiing in to the chocks just a short time before this incident. We ran a fuel usage study, recommending that hops be flown in the three-hour time frame, but the Group came out with the policy all hops would be only two and a half hours in duration unless you got into what helicopter pilots affectionately called a "shit sandwich." Then you could extend another hour. We changed our flight schedules accordingly, but most of us had faith that the little bird was good for at least three hours and more. She had two wing tanks. Most pilots flew one and a half hours on one tank, switched to the other tank until it coughed, then back to the first tank. (If you were flying down around 100 feet, that "cough" suddenly started hands flying around the cockpit, a thing we called a "pucker.") This new policy didn't sit too well with the some of the older hands who felt our hops should be at least three hours long, but I must say it was sure a lot easier on the old "kabungus."

Right after this fuel decision, I was flying out west of Khe Sanh. I'd been on the left tank for about one and a half hours. When I switched over to my right tank, the engine quit. (I was about 50 feet off the deck.) I quickly switched back and the engine caught. I climbed to altitude, tried again, and it quit. How much gas did I have left? Hell, who knew? Figuring I had enough to get home, I was greatly relieved when I arrived over Vandy Base. From here on home, I could always land on the road in friendly territory. I made it home, climbed up to check how much fuel was left, and found out I had improperly secured the gas cap to the right wing tank. All the gas had been sucked out into the wind stream. The left wing tank had four gallons left. No sweat. The next day I purposely ran one tank dry and timed it: three hours would do just fine, with enough fuel to comfortably fly back to Quang Tri from just about anywhere in our operating area. Leaning out the fuel mixture provided a lot of additional flight time, but due to a lot of maneuvering and altitude changes, most pilots were inclined to fly at full rich.

To understand this next event, I must explain two things: the leech problem in Vietnam and a venereal disease called non spec (non specific urethritis). Any veteran of NAM can tell you those little leeches were trouble. The leech a little bug similar to our American tick. He would attach to your body, bite into the skin, inject a fluid into the wound that caused your blood to ooze, and would then gorge himself on that blood. I've heard they can swell up to the size of your thumb before dropping off.

Non spec was a disease first discovered in the Far East. Unable to properly classify it in the venereal disease family, it was simply called non specific urethritis. Transmitted through sexual intercourse with local bar girls throughout the Far East, you knew you had it when you got up one morning and discovered you penis was rather swollen and you couldn't urinate, or the act was very painful. A shot of penicillin solved the problem, and the recipient went merrily on his way.

One day just about dusk, I received a call from our ground controller asking us to fly out to an area and check out a recon team to see if they were okay. On the way out, I heard the team radioman trying to contact what appeared to be his home base. His people came up on the air, and keeping in mind the recon teams were required to whisper all their transmissions, the conversation went something like this:

"Demon, this is Base reading you, go ahead with your request."

"Base, this is Demon, are you going to be able to lift us out of here before dark?"

"Demon, it doesn't look like we have choppers available. You'll probably be spending the night in the bush."

"Base, I've got a problem and need to be out of here, if at all possible."

"Demon, what's your problem?" (If a recon team stated they had a problem, they were probably getting into a tense situation and might end up in a "shit sandwich.") Demon surprised everybody on the frequency by whispering, "Base, I've got a man been bitten on the penis by a leech, and it's swollen up about twice the normal size!"

Base came up laughing in his mike and said, "Demon, repeat your last transmission." (Demon repeated the transmission, and I just knew South Vietnam was getting a real chuckle out of this one.)

Base, still laughing, came back with, "Demon, wait one, and I'll check it out for you."

Later, Base came back with, "Demon, this is Base. I've got a doctor on the line. Talk to him."

"Doc, this is Demon, I've got a man here that has been bitten on the penis by a leech, and it's swollen up bad. Can you get us out of here?"

"Demon, is the leech still attached?"

"No, sir, it has dropped off."

"Demon, he should be all right."

"Doc, I've got another problem, the man that got bit on the penis has non spec, and I really need to get him out before dark."

"My God . . . can he urinate?"

"Wait one." (A minute later.) "He can urinate, but it's sure painful."

"Demon, this is Doc, they tell me there is no possible way to get you out tonight. Give him some APCs, and I understand they are planning to get you out tomorrow morning."

As the conversation ended, I hit my mike button and asked, "Demon, this is Sea Worthy 22. Have you got that leech?"

Demon, came back with a rather quizzical tone to his voice, "No, sir."

With tongue in cheek, I said, "You'd better find that leech. We don't want him spreading non spec all over Vietnam!" Demon wasn't whispering but was laughing out loud when he responded.

Recon mission . . . I loved the recon guys. On Recon inserts, we would chopper over to 3rd Recon, plot the coordinates where each team would be dropped off, briefed for awhile, then they'd load up and head out. The gaggle usually consisted of two Huey gunships and a CH-46 or a couple of Huey Slicks as troop packers. I got to meet some of the recon troops and talk about their role in the war. We put these guys in some of the most ungodly and dangerous spots in Vietnam. Many were up on the DMZ and way out on the Laotian border. A team was usually made up of six to eight Marines. They were inserted into an area to reconnoiter for four or five days or until surrounded by the enemy and had to be heli-lifted out. The latter usually developed into "shit sandwiches"—many at night, under fire, which made it extremely "touchy." The teams usually sat up near an infiltration trail or overlooking an area frequented by enemy forces. They looked rather spooky with their camouflage-painted faces and always loaded down with weapons and communications paraphernalia. They took on an eerie demeanor.

These kids were our spies in the bush, and they did a hell of a job obtaining intelligence information. I talked to one team leader who said his team was moving down a trail out on the Laotian border when they heard movement up ahead. They only had time to slip three to four feet off the trail into the bush before the gooks came in sight. A platoon of enemy soldiers could easily reach out and touch any one them. There was also a story going around that one recon member in hiding was so well camouflaged that an enemy soldier had actually walked off a trail and urinated on his legs without the guy even seeing him!

During this stretch of time, we were warned to be on the lookout for a gay civilian. Seems this guy had fallen in love with a Major in the Corps, had followed him from stateside, and was hunting for him all over Vietnam. Apparently, he had falsely obtained orders to Vietnam, had stolen various uniforms, had been seen frequenting various O'Clubs, had been able to finagle orders to get from base to base, had slept in both officers' and enlisted quarters, changed rank from Sergeant up through Captain, and was rather doing his own thing to find his lover. The day the bulletin came out, some of our guys remembered a strange face in the O'Club the night before, and one of our S-1 clerks, while working late, had spotted an officer he didn't recognize sleeping out in the S-1 bunker. He had shaken him awake but let him stay after the guy pleaded drunk and lost. Don't know if they ever caught up with him.

An Unforgettable Character

Commanding Officer

Lieutenant Colonel H. A. (Fritz) Zander

(He was just a little ol' Major when I knew him)

Stand tall, you have served with the best.
Let us not forget our fallen comrades, and as we go forward
take as our theme the words of the song . . .

"And the world will be better for this,
That one man, scorned and covered with scars,
Still strove with his last ounce of courage."

Cast of Characters

Major M. L. Whitehouse

Major H.W. Shaw

Major K. R. Buske

Captain T. R. Irvine

Captain C. D. Saunders

Lt. S. T. Palmason

1st Lt. J. H. Lawrence

Captain R. H. Meydag

1st Lt. R. J. Moriarty

The O-1 is the oldest tactical aircraft in the Marine Corps. Its ability to fly low and slow over enemy territory to find targets is unequalled—a very vulnerable aircraft which is flown only by men of skill and courage.

Hey, you got any baling wire?

GySgt. G. B. Lockwood
O-1 Maintenance Chief

The Little Bigger Bird

The OV-10 Bronco was the Marine Corps' new light-observation aircraft. It was flown by VMO-6 pilots and carried Aerial Observers of the 3rd Marine Division. It has unexcelled visibility for observation and also the ability to protect ground troops who are in close contact by firing its own internal ordnance.

The Huey

The UH-1E is a versatile, turbine-powered helicopter. VMO-6 flew the "Gunship" version. This configuration had four forward-firing machine guns and carried two pods of 2.75-inch rockets. It also had two door-mounted M60 machine guns which were manned by gunners and crew chiefs.

Out of Chute #1

Majors' Hut

Buske—Combat Ready

General's Shower
(The Major's when the General wasn't around)

The Officers' 3 Holer

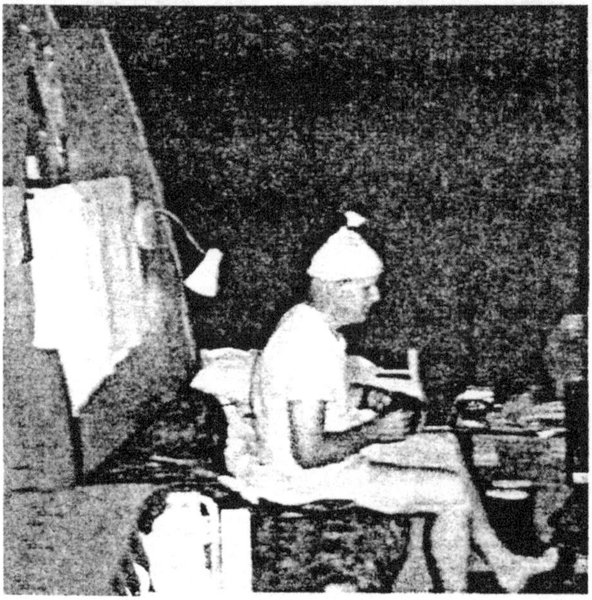

Fritz . . . To Bed

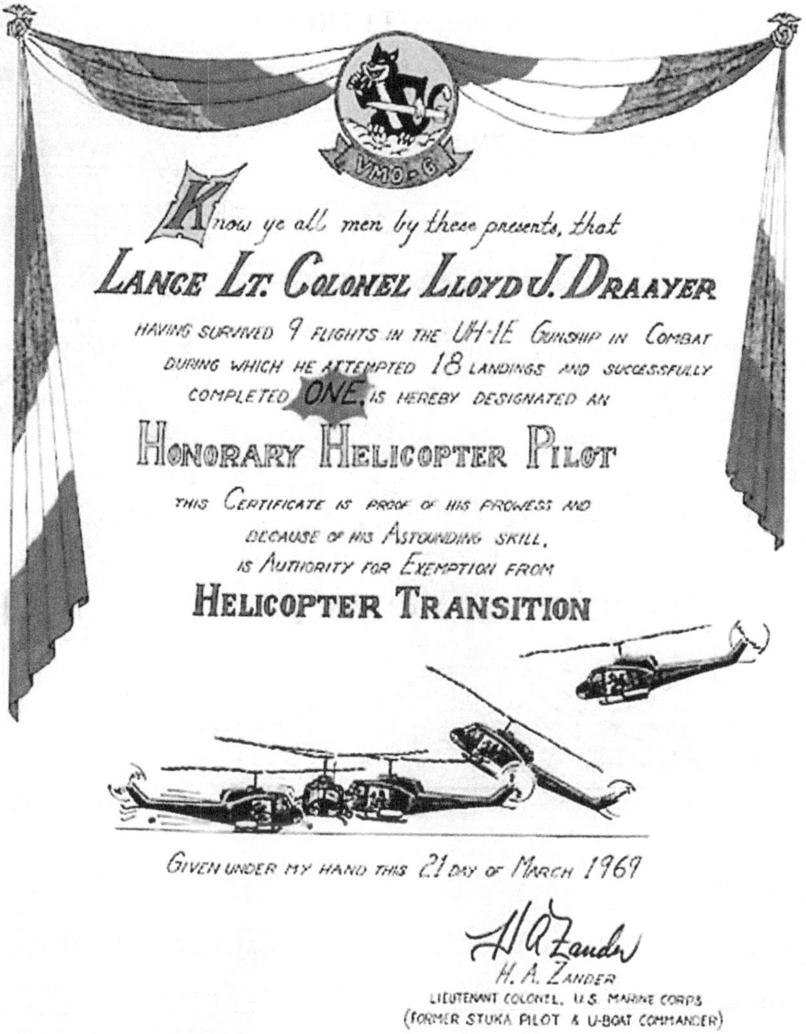

VMO-6

Know ye all men by these presents, that

LANCE LT. COLONEL LLOYD J. DRAAYER

HAVING SURVIVED 9 FLIGHTS IN THE UH-1E GUNSHIP IN COMBAT DURING WHICH HE ATTEMPTED 18 LANDINGS AND SUCCESSFULLY COMPLETED ONE, IS HEREBY DESIGNATED AN

HONORARY HELICOPTER PILOT

THIS CERTIFICATE IS PROOF OF HIS PROWESS AND BECAUSE OF HIS ASTOUNDING SKILL, IS AUTHORITY FOR EXEMPTION FROM

HELICOPTER TRANSITION

GIVEN UNDER MY HAND THIS 21 DAY OF MARCH 1969

H. A. Zander

H. A. ZANDER
LIEUTENANT COLONEL, U.S. MARINE CORPS
(FORMER STUKA PILOT & U-BOAT COMMANDER)

Douglas A-4 Skyhawk Scooter
The best little ol' bomb-dropping platform in I Corps
flown by the best darn "droppers."

Northern I CORPS

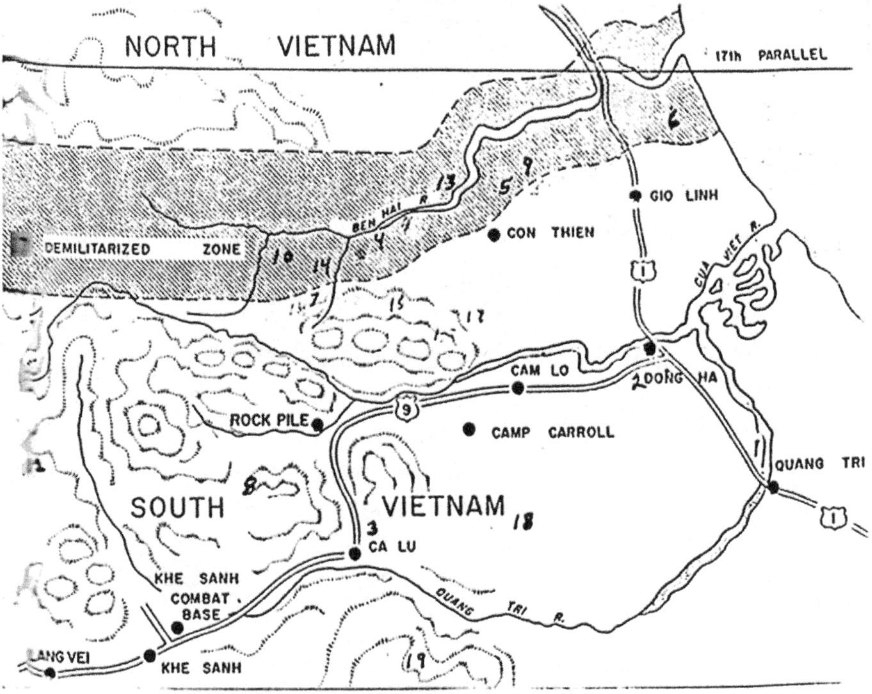

LEGEND

1. Quang Tri Air Base
2. Dong Ha Air Base
3. Vandegriff Combat Base
4. Submerged Rock Roadbed, Ropes Across Ben Hai
5. Giant Steps
6. VC In Sand Hole
7. Buske, M-16 Fire At Two VC
8. DFC Hop
9. Recon—Close Nape
10. Captured VC Lieutenant
11. Wassay Recon Insert
12. Whitey—Buffs in Crater
13. Alligator, Deer Cowboy Sightings
14. Elephants on Ben Hai
15. Naked Dead Marines
16. Huey Still Running
17. Last Two Hops In O-1
18. Nape in Football Stadium
19. Hit on Head
20. ??Hit in Laos

The A Shau

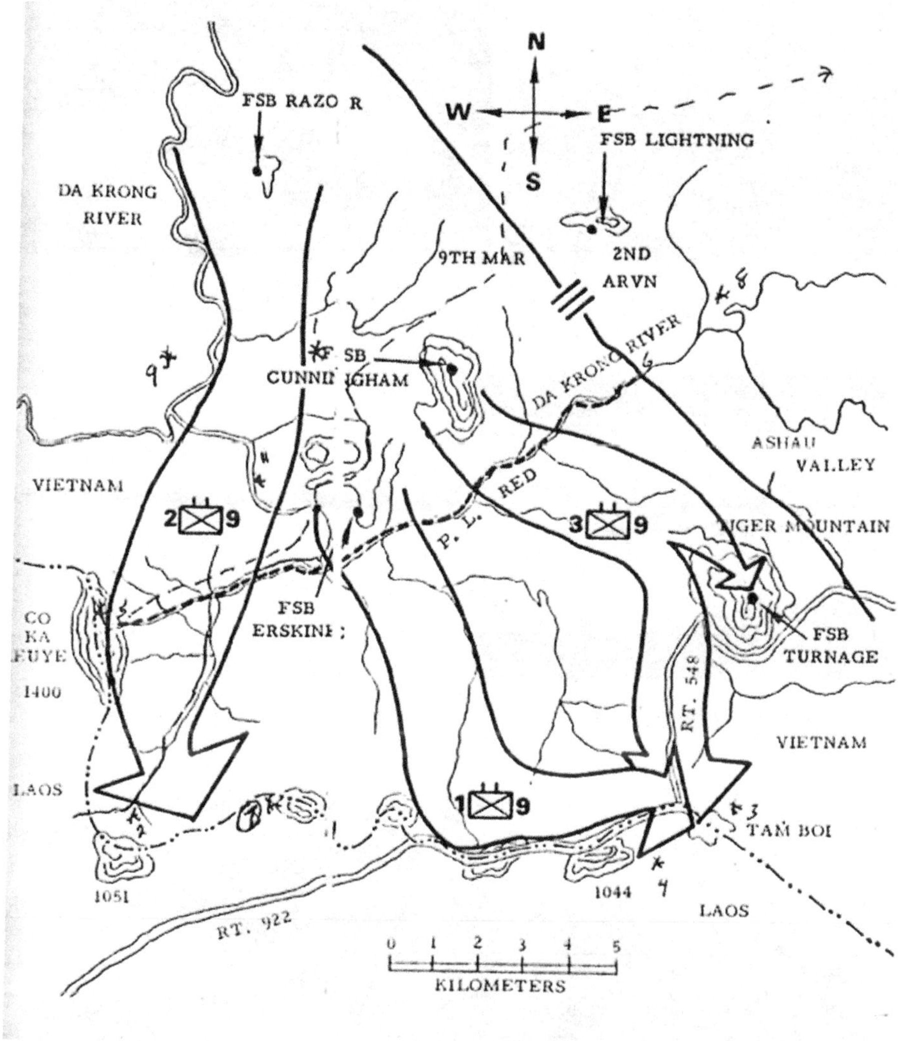

N
W
E
S
FSB RAZO R
FSB LIGHTNING
DA KRONG
RIVER
9TH MAR
2ND
ARVN
FSB
CUNNINGHAM
DA KRONG RIVER
ASHAU
VALLEY
VIETNAM
P. L. RED
2 9
3 9
TIGER MOUNTAIN
CO
KA
EUYE
1400
FSB
ERSKINE
FSB
TURNAGE
RT. 548
VIETNAM
LAOS
1 9
TAM BOI
1051
1044
LAOS
RT. 922
0 1 2 3 4 5
KILOMETERS

Khe Sanh After The Siege

FSB Cunningham

Before The Flood

During The Flood

Typical LZ

Fire Support Base

Early photo of Marine helicopter making a one-wheel landing on the Rockpile.

Later photo: Air knocked off the peak and Marines installed a Marston matting "runway," just large enough to accommodate one chopper.

Ville Irrigation System

New Concrete Well
(center by lady in white)

Cutting 2x4s

Grazing Water Buffalo

The USS Independence

"Let Freedom Ring"

February 2, 1966 – Completed five traps and five cat shots

The Frigate USS McDonough

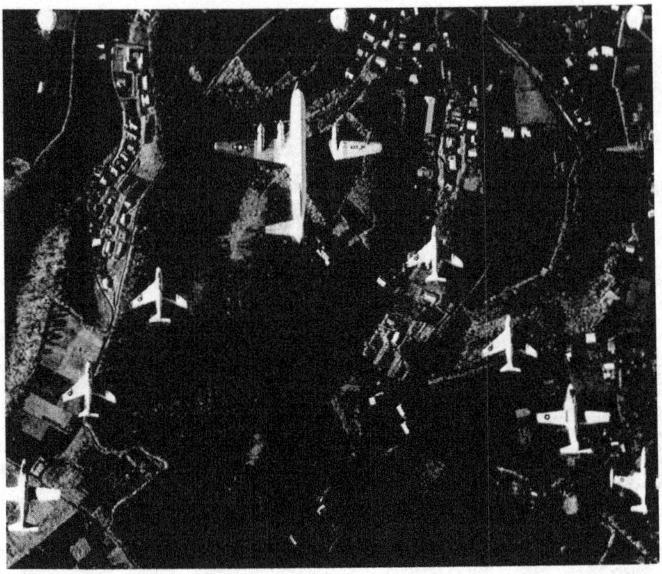

Secretary of Navy's Plane – Atsugi

Chapter 7

Camp Pendleton

To digress a little, all my experiences with helicopter pilots came during my "Grunt Tour" at Camp Pendleton. The Marine Corps seems to think a jet jockey ought to get down and grunt with the ground-huggers (ground Marines, or affectionately, grunts) sometime during his career. Every year, Headquarters Marine Corps selects aviators to serve with the grunts as FACs or Air Liaison Officers (a ground job). I fought my assignment but ended up at Camp Pendleton, California, with the 1st Battalion, 7th Marines . . . loving (almost) every minute of it. My experiences with the helicopter fraternity during my grunt tour left me with many misgivings concerning their professional capabilities. As I settled into VMO-6, I could see being co-located with a bunch of helicopter pilots might be a downer. My theory:

Sometime during the early 60s, the USMC was greatly expanding its helicopter capability. Not having enough helicopter pilots coming out of the Training Command pipeline to fill the need, they went to the fighter/attack squadrons for jet pilots. Now, no fighter/attack jet jock wanted to volunteer for helicopter training, so it boiled down to demanding a quota from each jet squadron in the Marine Corps. Naturally, the Squadron CO wasn't going to sacrifice his best pilots to this solicitation, so the choppers got the pick of the worst jet jocks. Don't get me wrong, some damn fine pilots ended up in choppers, but in general, the chopper people got the bottom of the bunch.

My first encounter with chopper pilots came one day when several FACs were sitting on a hill waiting for four choppers to arrive from the Marine Air Facility up near El Toro (the field with the big blimp hangars). They eventually arrived, late as usual, and when we met to brief the day's work, it was discovered they hadn't brought the appropriate maps. To resolve the situation, one pilot, a former jet jockey and a hot one at that, volunteered to fly to the Camp Pendleton airfield to obtain the maps. His chopper was sitting behind the flight leader's helicopter. He revved that rascal up, took off straight up and then back in a rather violent maneuver, one I had never seen before. I took a quick glance at the other seven chopper pilots. They were standing there in awe, eyes as big as saucers, not believing what they had just seen. When the jet jock returned, I asked about the maneuver. He stated, "Hell, it's a simple thing but very few chopper guys have the balls to try it."

Chopper drivers were lousy navigators, never carried the appropriate maps, and I hate to say it, frequently got lost. One incident comes to mind. One senior chopper driver's flight of four took on a bunch of our troops at Camp Pulgas with instructions to debark them at a prominent landing site three or four miles to the east. His first attempt landed our troops in the highly restricted ammo dump near the Fallbrook back gate. This ammo dump was so highly classified and protected the MP sentries even timed automobile traffic traveling across the area to discourage slowing down or stopping. One of our squared-away Sergeants immediately recognized the mistake and talked the pilot into lifting off before the troops were disgorged (and before the MPs started shooting). The next landing happened to be about two miles off the base. Once on the ground and realizing his mistake, the troop leader decided to get to a highway and thumb back to the Fallbrook gate. As they were searching for a road, they entered a lady's backyard. She looked up from hanging her laundry and exclaimed, "My God, the Marines have landed—I'm being invaded." They used the lady's phone to call the Battalion for some truck transportation.

Our Battalion CO was a terrific guy who loved to use air and let us schedule it probably more than any battalion on the base. One such opportunity came up during a planned exercise in the hills just west of Camp Pulgas (a big order): land our troops out of UH-34 helicopters into twelve different landing zones. The difficulty going into a landing zone for the ground troops revolves around being dropped exactly on the pre-selected site. If this occurs, the troop leader can quickly orient his position and immediately assume attack headings, etc. If he's dropped away from the expected site, it takes time to sort things out. Generally speaking, a troop leader will be dropped into a completely unfamiliar site anyway, so if he's dropped at a spot some distance from his intended site, he's in a void. In those days, helicopter pilots didn't seem to appreciate this gentleman's problem.

Being a photo pilot almost all of my flying years and knowing our troops were probably going to be dropped off at the wrong landing sites, I requested forward firing photo coverage into each site from my old photo squadron stationed at El Toro. A few days before we were to kick off the operation, I hadn't received the requested photos, so I finagled a chopper ride in one of our little Kaman "Huk" helicopters located at the Camp Pendleton airstrip. Armed with a "newspaper"-type hand-held camera, I flew down the coast from El Toro, took pictures along the route at each turning point, then the track to be flown into each landing site, and a photo of the exact point of landing. We ran low on fuel before I completed my photo work, so I missed getting three sites. In essence, I was giving the helo-pilots a forward-looking view of their entire route.

First Lt. Mike Whitehouse, my companion FAC in the Battalion, and I sat down, traced each approach on maps, and forwarded the bundle along with the photos to the helicopter MAG (Marine Aircraft Group) supporting the exercise and felt it was very well planned. However, as a safeguard, I took a day off and drove up to the MAG. I was actually able to sit down and brief with the helo-Squadron CO and most of the pilots who would be flying our troops. They had a little trouble finding our bundle but, when they did, we oriented the photographs with the line of flight on the map, and they were tickled to death with the photos. It was going to be a piece of cake.

It was . . . almost. The choppers dropped their loads exactly on the nine sites that had photo coverage. The other three flights, without photos, dropped our troops some distance from their landing sites. They subsequently became disoriented and were a couple of hours tramping to the proper site. Although I was a little disappointed with my old photo squadron for not delivering the goods, the results are absolutely amazing when the full capabilities of the Corps are cranked in while planning an operation. It takes complete ground and air cooperation.

This is an area that completely mystifies me about the Marine Corps—the traditional feeling of hatred between the grunts and airdales. (We were called airdales or zoomies by the ground people.) Aviators, especially in the senior ranks, seem to feel they are superior to their counterparts on the ground. (Grunt is not a derogatory word in my vocabulary. I loved working with them, and, except for one incident, never got any static from grunt land, but I sure did from some of the helo-people with their stubborn superiority complex.)

My first incident occurred during a scheduled heli-lift out of the San Onofre area. Our troops and equipment were at the dirt landing strip waiting to embark. The choppers were to set down, take on the troops and equipment, and then land them at another designated site. Simple—never have to shut down the choppers and, after delivery, go home. The choppers landed and then shut down. The Flight Commander, a Bird Colonel, got out of the lead chopper, walked over to my FAC radio jeep, and stated, "I will not take that jeep. It is too heavy. I won't take that one either, nor that one. They are all over weight. Haven't you weighed them?" I was flabbergasted. Every good Marine knows that every jeep, piece of equipment, water can, etc., has its weight catalogued or stenciled somewhere to specify the weight of the item. As always, Mike and I meticulously checked over our figures and knew the loads were within heli-lift requirements. The Colonel stalked off, mumbling back to his chopper. Finally, a foreign exchange pilot with an English accent pulled out his calculator, figured the weight from our stenciled equipment, temperature, humidity, etc., and said, “I believe we can lift your equipment.” He strode over to the good Colonel, some words were exchanged, the pilots all got back in their choppers, and the lift went smooth as a baby’s bottom.

This Colonel soon got his put-down, thanks to our Regimental Commander Bird Colonel (Acid Al Arsenalt, 2nd Battalion), as a climax to an exercise that had been going on for about two weeks. He had scheduled a large Battalion lift from the Pulgas area to an LZ on top of K Springs Mountain. Mike was called upon to be at the pickup site to relieve 2nd Battalion FACs going on the lift. The lift was scheduled for 0800 hours. The troops were still eating or cleaning mess gear at 0630, when Mike looked up and observed a gaggle of choppers approaching. They landed. Lead (the aforementioned Bird) came up on the radio and stated, “We are here—load the troops.” Mike hit his mike button and said, “Sir, the heli-lift is scheduled for 0800. You are too early.” Lead came back with, “I’m not shutting down these helicopters. Load those troops. If they are not loaded in five minutes, I will take off without them.” Mike was aghast and didn’t know what to answer. Suddenly someone came up behind him and asked for the mike. It was Acid Al. He had been standing behind Mike and had overheard the conversation. Now, Acid Al was a great soldier, had great respect throughout the regiment, and was as tough as his name implied. God bless him, his first transmission was, “This is the Regimental Commander speaking. I don’t know who is on the other end of this line, but shut down your helicopters and meet me in the tree line off to your right!” Mike said he felt like cheering. He said the choppers wound down, one by one, and the good Colonel stepped down to meet his opponent. The two “Birds” met. There was some lively discussion, but somehow Colonel Arsenalt knew he had the numbers and was senior to the intrepid aviator. As a result, the chopper pilots twiddled their thumbs until 0800 . . . and not a second sooner. God, I loved that man. We heard a rumor Colonel Arsenalt had threatened his opponent with putting him on report if he didn't cooperate.

The only bad incident I had with the grunts involved my being a rodeo clown. I was actively clowning professional rodeos in the Southern California area. In fact, I had just recently clowned the annual Camp Pendleton rodeo. When we weren't working on weekends, I could clown on Saturday and Sunday and make $200, a nice addition to a Captain's pay. One Monday I received a call from a clown acquaintance, stating a rodeo up in Canada had their scheduled clown hospitalized by a fighting bull. They needed a replacement that weekend, and the job would be worth $600 if I could make it. I wasn't really busy at the time so went to my immediate boss, the S-3 (Operations Officer), a Major, for permission to take seven days leave. He would not approve the request, leaving me rather frustrated. Two weeks later he took leave.

While the Major was on leave, his assistant (a Captain) and I were sitting there staring at each other, trying to think up something we could schedule for the troops. I had recently read an article in the Base paper about some Recon troops rappelling out of choppers. I suggested we do that with all the troops in the Battalion. He liked the idea, and with a "cat gotcha grin" on his face said, "Jeez, the S-3 would never let us do anything like that. But he's on leave, so let's do it." We obtained . . . someone's approval . . . I think . . . so I requested the choppers.

We didn't have any rappelling equipment in the Battalion, so I drove over to Recon to borrow some. Neither did Recon. (The fancy rappelling hooks and paraphernalia in existence today were not in the Marine Corps inventory yet.) I was briefed on the procedure, picked up some 200-foot-long ropes from supply, and headed back to Pulgas. We rounded up my FAC team, went over to the camp swimming pool, chose one of the tallest diving towers, named it the Pulgas Helicopter, and practiced the rappelling procedure.

Rappelling from a helicopter in those days was not an exact science, but the existing procedure of standing or sitting in the chopper door, getting a death grip on the rope above the head with one hand, wrapping it around the leg, down under the instep, then up through the free hand, and then stepping vertically out into space seemed adequate. All one had to do was feed the rope through the free hand to let yourself down. It seemed to work. We tried it—me first, because up to this point I was the expert, then the Assistant, followed by my FAC team. It worked fine.

The next week, we mustered all the troops available in the Battalion at the swimming pool and proceeded to teach them to rappel. Being Marines, most of the Battalion troops picked up the trick easily. All they had to do was string the rope properly and step off a platform into thin air. Our helicopter was probably forty feet from the ground, so a few of the troops had to be goaded into pushing off from the platform. Their faces showed a great sigh of relief once they safely established their weight on the rope. We had a couple who absolutely refused to go through the training. They were simply afraid of heights, so were excused; but many of the troops enjoyed rappelling and went down the rope several times. The day the choppers were finally scheduled, more than half the Battalion troops were away from camp, but some 250 bodies showed up. That was okay—it would have taken us two or three days to get the whole Battalion up and out of the choppers.

When the choppers arrived, we shut them down for a briefing. Lead was a Major who didn't especially like the idea because no one in his flight had ever done anything like this before. He didn't particularly like the idea of a rope hanging out and under his chopper, feeling it could sweep up and become entangled in his rotor blades. We volunteered to hang onto the rope as he lifted off and then hang on to the end of it from underneath the chopper while he hovered about tall, tall, telephone pole height above the ground. He didn't like having people beneath his bird, but we convinced him we were pretty fast runners. We tied one end to the chopper boom, and with me playing out the slack as he took off, the rope problem was solved. Everything went well, so we started the lift. The troops just loved it. Most of them wanted to go more than once, which time wouldn't allow.

Three incidents come to mind:

Some of the troops, once they got the feel, really sped up about half-way down the rope. Others, less confident, came down slowly. One individual slipped into a horizontal position and gently lowered himself to the ground on his back. (He won the funniest descent award.)

One guy lost his instep hold and hung entirely by his hands. He eventually got his legs around the rope to take off some of the weight and let himself down. (He was voted the man in the Battalion with the strongest grip . . . and the ugliest rope burns.)

Toward the end of the day, one of the troops must have forgotten to distinguish his left hand from his right. Once out of the chopper, his brain told his right hand to loosen its grip so he could start his descent. His left hand answered the call, resulting in an immediate transition to the head-down, vertical position. He must have then decided that was the wrong thing to do and let go with his other hand. Lucky for all of us, the rope fell into a half-hitch around his ankle and, without anything else touching the rope, he probably made the nicest, most spectacular, slow rappel of the day. Man, was I relieved. (He won the Left Hand/Right Hand Award.)

I think the chopper pilots actually enjoyed the day. They got in a lot of hovering practice. We shut them down for lunch and treated them with a visit to our mess hall. Bless their little ol' hearts, I would have hated to fly all day over the same spot on the ground. Word quickly spread amongst the returning troops that night, and we were constantly asked to reschedule the choppers. Most were really put out they didn't get a chance to participate. The Assistant knew the Major wouldn't allow a repeat performance, so it was forgotten. Unbeknownst to us, a Base reporter attended our session, took pictures, and did a rather nice article on our little enterprise. Someone, presumably an enemy, placed a copy on the Major's desk the day he came back. He was livid, went straight to the CO, and that's the last time anyone ever mentioned it again. We had won! This little trip was one of the greatest confidence builders in our Grunts I observed during my tour, and to quote some of them, "Man, I'm writing home about this one!"

A short time before I was scheduled to return to the flying world, the 7th Marines conducted a large desert exercise out at Twentynine Palms. In those days, ground communications were practically nil, and after the second day, our Battalion couldn't talk to a soul except by relaying message traffic through the O-1 Bird Dogs on the UHF/FM radios. Apparently, our unit was the only one using O-1s, so we had plenty to work with. They were just great! Finding the enemy aggressors was a snap in the desert for the Birddoggers, and without air available, we used them to simulate attack aircraft by dropping small paper sacks filled with white flour on the aggressor positions.

The day before the operation ended, we asked the Birddoggers to drop us the equipment needed for a message pickup. We received the drop about noon and set it up. Simple: stick two ten-foot-high poles in the ground, some ten feet apart, string a small rope from pole to pole, tie an empty hand grenade canister to the tail of the rope, and as the pilot flew the little bird over the layout, the backseater reached out the window with a hook, snagged the rope, and hauled the canister aboard. (Sounds ancient, but by golly it worked.)

While my FAC Team set up the contraption, I slipped thirty bucks into the canister and a note thanking them for all the good work they had done for us. I told them the Frosties would be on us at their "last night, going-away party." (Frosty was the Birddoggers secret code word for a cold beer. They knew we grunts in the field were denied alcoholic beverages . . . thus the secret word. Original, huh?)

About noon the next day, the S-3 Major was near my radio jeep, as he had been throughout the exercise. A Bird Dog flew over and, bless his little heart, called, saying . . . "Good afternoon, Blazer, I have message traffic for your S-3 . . . and a Ffrrossttieee for you." I was in trouble! I told my FAC team to run out and gather up the canisters as soon as the Bird Dog dropped and run like hell over to my jeep.

Just about the time of the drop, the Major finished recording the messages and then asked me to explain the meaning of Ffrrossttieee. As I was trying to think up some brilliant answer, my FAC Team ran up with six canisters. I grabbed the first one, handed it to the Major and said, "Open it up and you'll see!" He was a little flabbergasted when he opened it up to find a cool can of beer. He set it on the hood of my jeep, and when I told him it was from the Birddoggers to him, he reluctantly pulled back on the tab. It squirted everyone in sight, especially the Major. The beer had been dropped from about ten feet, sans parachute, and tumbled a couple off about 100 feet along the ground . . . so you can see . . . it was ALL SHOOK UP. The Major was a good sport, took a couple of sips, and passed it around to the troops.

What did I feel about helicopter pilots in Vietnam? Some of the best I ever met. These guys continually went out on shit sandwiches, won many high decorations, deserved every one of them and probably more. Their missions entailed flying at night in some of the ruggedest terrain in the world, in all kinds of weather, and I never saw one of them flinch. A great bunch of guys, professional in every respect, and fun, fun, fun! I was extremely proud to call them squadron mates.

Chapter 8

Giant Steps

I had a pair of flight boots that I dearly loved. I'd had them for years, but I checked out a pair of cloth Vietnam boots, thinking they would be cooler in the summer. I was now flying over enemy territory every day, and if rudely shot down and I had to hike home, the new ones were thought to be the "in thing" for jungle walking. The day I made the switch, I took them with me on my morning hop, thinking, "Hell, I couldn't carry any bombs if I got the chance—why not try a boot?" Can't you just see some gook's family getting a letter from Hanoi stating their family member had just been KBB (Killed By Boot)? Never did get the opportunity.

We received our first OV-10 Broncos into the squadron November 1. It was a beautiful bird, and I couldn't wait to get my hands on it. The OV-10 pilots were a hard-charging outfit, itching to get into the fight. Fritz wanted to check out in the bird and invited me to do likewise. I was finally going to be flying a seat that could shoot back.

The next day found me in the Bird Dog flying in an area just on the northern edge of the A Shau Valley. I had a big cannon coker in the back seat, WO Luckenbaugh, an excellent AO. I spotted a large, thatched hut just inside the valley. After receiving clearance to hit the target, two Air Force F-4s showed up carrying 750 pounders (large bombs). Lead's first hit was a little short; Two's were right on. As Lead came around again, I was, as usual, paralleling his run-in heading. As he dropped his bombs, I started my off-target left turn. I heard a loud explosion back at my eight o'clock. Just then something hit my aircraft, Plexiglas came raining down, something hit me in the shoulder, and the aircraft started shaking violently. It was mayday time!

I knew we had taken a pretty good hit, the engine was running rough, and we were headed for the trees. I called the F-4s, stating my predicament, and asked for cover if we went in. There wasn't a clearing in sight to land the little bird, so it looked like we were going to test the treetop landing theory. Lead came back with, "Can we finish our runs to expend our ordnance?" Here I'm headed for the trees and this guy wants to drop some bombs that might be needed to help me out later. I gave him an ass chewin' to get him back into the real world, found the vibration had subsided somewhat, turned toward Quang Tri, and started a slow climb out of the valley.

The F-4 driver went to altitude, contacted control, briefed my situation, and then stated that, due to low fuel, he had to leave the area. (I was too low for my mayday transmission to be heard.) I jumped his butt and told him to get me someone to fly escort on us to monitor our trip back to Quang Tri, if we could make it that far! I sure had a lonesome feeling in my gut watching the F-4 flight disappear into space.

While checking the O-1, I noticed the top window had been broken out. I looked back at Luckenbaugh and found him with a blank, uncommunicative stare on his face, a broken visor and blood coming from a cut on his chin. (For some reason, he was the only AO who frequently flew with his helmet visor down.) He wasn't going to bleed to death but appeared to be in shock. We reassured each other we were both okay and headed for Quang Tri. About halfway home, two CH-46s joined up. Just to show off and let them know where my problem was, I stuck my hand up through the broken window, waved, and then gave them a thumbs up. They detached when we were over Quang Tri. The landing was uneventful.

As we exited the plane, we conducted a little cockpit search. I found a large chunk of metal lying under the back seat. It was about 3/4 of an inch thick, 7 inches long, and over 2 inches wide. Tracing its trajectory, the chunk hit the metal frame of the top window, leaving a considerable dent, exploded through the Plexiglas window, ricocheted off the front of my shoulder, and still with some force tumbled back and hit Luckenbaugh in the visor. The jagged metal mass pierced the visor and then hit him on the chin, drawing blood. He didn't want to, but I insisted he go to sickbay to get it doctored. I figured he had been through enough to deserve a Purple Heart for the blood running down his chin. He got his Purple Heart. I looked under my flight suit at my shoulder to discover only a large bruise starting to form. I jokingly told Luckenbaugh, from that point on I was always going to carry a pocket knife in case I got hit by a flying projectile. If it didn't draw blood, I intended to stab the thing to draw some for my Purple Heart.

In discussing the incident a little later, I mentioned the fact that I couldn't figure out why the aircraft was shaking so badly. His reply was, "That wasn't the aircraft shaking, Sir, that was me . . . shaking the aircraft! When we got hit, I was just so scared my whole body was shaking." S-2 sent the chunk to experts in Saigon to analyze. They told us later the thing was from a VCRPG 7 (Rocket Propelled Grenade, anti-tank missile). During my FAC tour, I heard the same loud report several times. It seems the gooks thought they could hit an aircraft with the weapon and, although hard to believe, we later learned the little rascals would try. I had a sneaking suspicion, however, that I had been flying a little close to the target for 750 pounders and was my own BDA!

Major XX (name withheld), a chopper pilot, was a rather excitable individual. This was brought out forcefully one night as we Majors were blissfully sleeping. All of a sudden, Major XX burst in the back door screaming, "They are at the wire attacking Quang Tri. Get your gear and follow me. Come on you guys, we've got to defend the Base!" Groggily, we wondered what in the hell had got into ol' XX, as he tore out the front door on the dead run in full fighting gear. We walked out to see what was going on and heard a couple of loud explosions just north of the Base. Another roomy came out of the hut about then and mentioned that Major XX had been over at the club tossing a few when someone picked up the duty microphone and announced that the gooks were about to attack Quang Tri. Major XX had taken the announcement at face value and was about to give his all for Quang Tri. We believe he remained out in the trenches until dawn. (We found out the next day that the three explosions were Army arty rounds mistakenly fired onto Quang Tri—no damage.)

Each Major in the Majors' hut had a little 10 x 15-foot cubicle bordered by lockers and open to the middle of the Quonset. Fritz's and mine were located at the front door. The remaining Majors spread back to the kitchen, with Major XX occupying the last bunk on the right.

One event we watched with great humor was Fritz's nocturnal visits to our outdoor piss tube. (The piss tube was a simple, four-inch diameter, white plastic pipe, sunk in the ground and tilted at an angle to match the incoming flow, which extended from ground level to about "the right height." They were spread out all over the base and were usually out in the open). There were nights Fritz didn't make it. One late night, two roomies were sitting in the kitchen when they saw Fritz walk slowly down the hall in his skivvies and stocking cap, stop, lean up against the wall—and pee on the foot of Major XX's bed. The laughing observers tried to wake him up, but Fritz staggered back to his cubicle and was soon fast asleep. When queried the next morning, Fritz couldn't even remember getting out of bed.

For the next couple of weeks, every time Fritz got drunk at the O'Club, we kept a late night Fritz pee watch. One night I was reading in my rack when Fritz stripped down to his skivvies, put on his stocking cap, and went to bed. A couple of hours later, he stirred, got up, and I swear without opening his eyes started down the hallway. I asked him where he was headed. No answer. I hollered out, "Hey guys, watch out, Fritz is headed for the piss tube." Most of them woke up and looked on in amazement as Fritz staggered up to Major XX's bunk . . . and peed on it. Major XX was asleep in his rack, woke up, and screamed at Fritz. Not responding, Fritz simply made a closed-eyes wobble back to bed, not remembering one thing about the incident the next morning. There were two more trips, but try as we might, we physically could not steer him in other directions—he always ended up at the foot of Major XX's bed. The offended Major was at wit's end and threatened to shoot Fritz if it happened again. Just as suddenly as it started, it ended. Thereafter, Fritz made it to the pisser every time nature called.

The next incident was my DFC hop on November 17. W/O Luckenbaugh was in the back seat, and as we were merrily tooling around, we received an "in the blind" call from a recon team that was in trouble. We quickly switched to their frequency and found out we were within two miles of their position. I put in a request for air, and Luckenbaugh contacted arty. When over the team's position, I could see them throwing hand grenades and the impact of the grenades in the tall buffalo grass to their immediate front. The team was located in one of two large bomb craters. I immediately dove and fired a rocket at the intruders, hoping to take some pressure off the team. Would you believe, neither air nor arty were available. While continually alternating firing runs with dummy runs, I went out on Guard frequency, praying that some hard steel was in the area and could help out. Luckily, we were carrying six rockets that day, and just as I got down to the last one, a beautiful ol' OV-10 came around the hill. It was Captain Doug Shore from our own squadron. He joined me in the bombing pattern and the team soon reported the firing had slackened.

The team called several times during our runs, saying they were still taking fire, but the gooks shifted their sights to our diving aircraft on every pass. We could hear a lot of gunfire and assumed we were taking our share. I remember one call in particular. The team called, saying, "You guys have got to be crazy—you're taking fire on every run." We were rather busy at the time, so let it slide. It wasn't very long until Captain Shore called saying he was getting low on ammo. He had fired all his rockets and was about to start using his mini-guns. In the Bird Dog, we were out of rockets and had reverted to dropping smoke grenades on the enemy position. When we were out of grenades, I told Luckenbaugh to un-limber his M-16 rifle and start shooting on each pass. I further advised Shore to make a few dummy runs to keep enemy heads down. After several more passes, the team reported the firing had stopped. (I had moved the team out of harm's way, back to the second bomb crater. We later spotted a body in the first bomb crater. Don't know who got him, but I'd like to believe he was hit on the head by one of Luckenbaugh's smoke grenades.)

I had been making treetop-level runs, right over the gooks' heads, and will never know why we didn't get hit. Neither did the team. During these, as the chopper pilots called them, "shit sandwiches," you don't even think about getting hit. Most of the pilots I've ever talked to say the same thing. It's going to happen to the other guy . . . not me! It is just a matter of taking care of the present situation and not even thinking about the consequences. (Did you ever talk to any Marine who didn't think he was invincible?!)

Just as we were about out of ammo, two Hueys and a CH-46 called, saying they were inbound and had our position in sight . . . professionals had arrived (squadron mates). Major Dub Shaw, who was leading the flight, took over and made firing runs until satisfied it was safe enough to bring in the CH-46 to load the team. The pickup was a beautiful example of helicopter pilot professionalism. West of the second bomb crater was a small gully. The CH-46 pilot flew below into the gully to mask the chopper from any ground fire. Once up the gully, he did a 180-degree turn, backed up to the second bomb crater, dropped his ramp, and extracted the team. As he was pulling out, still down in the gully, I observed tracers flying near his aircraft—a damn gutsy maneuver.

I felt Shore and the CH-46 driver deserved some sort of decoration for their bravery, so I put them both up for DFCs. Luckenbaugh asked, "What about us?" Captain Shore also wanted to put us up for DFCs, and we probably deserved them, but I felt the

Medal Board might perceive our action as collusion just to garner a medal. (You know, I'll scratch your back if you scratch mine.) Additionally, I wasn't in this war to win medals, so I negated his recommendation. I figured with the dangerous flying I would encounter during my Bird Dog tour, I would certainly come up with a few medals. Most of our flights were of the lonesome variety and, be assured, most of the O-1 "daring do" was never observed by "our side."

This was my first helo extract, and I didn't realize until a later hop that the tracers I observed near the CH-46 were actually rounds coming from the bird's door gunners. Also, the CH-46 pilot didn't feel the flight was worthy of an award and told his superiors to cancel the recommendation—a man after my own heart!

During November, we had another one of our infamous cease fires. Orders came down to fly our regular hops but no shooting, bombing, or aggressive work in the DMZ. If shot at from across the river, retaliate, but only if they shot first. Apparently the gooks were going along with the idea. The next couple of days, we could fly along the Ben Hai with immunity.

During this stand-down, we had a visit from Liz Trotta, a TV reporter and her crew. She seemed interested in the FAC mission, so Fritz turned her over to me. To say the least, I was leery of the press. I had seen so much unfavorable comment and twisting of the facts by our news media that I was going to give her a story that would be the truth, with no room for exaggeration. I took her outside of our S-2 hut . . . forgetting the location of our outdoor piss tube. This one was located out in the open about fifteen yards from the back door of the S-2 shop, completely void of foliage or camouflage. I had my back to the tube and during the interview and noticed she was looking over my shoulder, apparently staring at the horizon. I turned to look and, sure enough, there was some Marine standing there urinating into the thing. Luckily he had his back to us. Liz didn't bat an eye, and I don't believe she even noticed the incident. I could see this was going to be a ticklish situation, so I maneuvered her around to place her back to the site.

Suddenly, our S-2 came out the back door and said, "Major, Captain Irvine just got shot up over the Ben Hai. They are running air on the guns right now." I excused myself from the interview saying, "Please stay right here. I've got to take a look at this." I went in where our people were plotting the position on our huge wall map and was being briefed on the situation when I looked up and there was ol' Liz, wide-eyed, notebook in hand, looking over my shoulder, cameraman rolling the film (with . . . "**This is a classified area, no admittance**" signs posted all over the room). I told our S-2 people to get her the hell out of there, stating this was a classified area. She said, "That's all right, we won't get in your way." We quickly booted her cute little butt out.

Soon, Captain Irvine came back with several holes in his O-1. Fritz and I debriefed him. He was on our side of the river, had taken some hits from their side, heroically stayed on station, and put arty and one flight of A-4s on the site. He got even! This was a clear case of the gooks breaking the rules of the cease fire. After approval from higher up, Fritz let Liz in on the thing so she had her scoop.

She left our area then and went over to 3rd Recon. About an hour later, she came back, escorted by the little Lt. Col. Commanding Officer of the outfit. He hit Fritz up with the idea of taking Liz and her crew up in a Huey, flying her over Khe Sanh, and a few other locations where his recon teams operated. Fritz agreed, admonishing him to not take them into any dangerous area. When they came back, their 1st Lt. Huey driver was appalled. He said, "We flew around the area. He wanted me to fly near the spot Irvine was just hit—I refused. We then went up to Khe Sanh. He told me to land and let them out. I refused. He insisted, so I landed. Now Khe Sanh hasn't been a hot area for months, but I felt really leery about the whole situation. It seems the whole idea was to let that reporter lady step out and get her picture taken with Khe Sanh in the background . . . just to say she had been at

Khe Sanh."

Supposedly, during this cease fire, both sides had agreed to cease maneuvering troops and supplies through the DMZ. Someone came up with a brilliant scenario: send some recon teams into the DMZ, set up a trap to silently ambush and kidnap the gooks who were infiltrating through the DMZ, take them back to headquarters, and conduct an interrogation to prove the NVN disregard for the cease fire. If a recon team was lucky enough to find and capture some gooks, a reaction force was kept in readiness, specifically to be heli-lifted in on a moment's notice to protect the team and their captives. Our mission was to fly around the area, to stay in radio contact with the recon teams, and ensure their safety.

The first morning of the operation on November 14, my back seat and I had visually located a team right next to two huge bomb craters. (We never flew right over a recon team's position nor dwelt in the area as that might give the gooks an idea where the team was located.) Suddenly, we heard gunfire from the team's position. We banked back toward them, found the team establishing a defensive position in the eastern of two bomb craters about thirty feet apart. We could see the impact of the grenades they were lobbing toward the enemy. An arty box had been pre-surveyed around the team. (An arty box formed a box-shaped pattern around friendly troops. If they got into trouble, arty could fire, setting up a steel wall of exploding shells protecting those inside the box.) While my backseat called for arty, I figured the gooks were in such close proximity to the team (inside the box) that artillery might not be effective, so I called for air. The team was experiencing a lot of pressure and needed help fast. The coordination worked beautifully. Two A-4s with nape and snakes, perfect weapons for this target, were already on their way. The birds must have been orbiting right over Dong Ha—and right here I have to regress a little.

During my grunt tour, I personally ran and observed many

napalm drops. To me, it was a weapon you could really bring in close to the friendlies, if they kept their heads down, and when dropped by A-4 drivers—accurate! Contrary to the malcontents in the states who were blasting it, napalm was the most effective, flexible weapon we used south of the DMZ, and it saved many American lives. Ask any VC who had a tank dropped on his head. There are not many of them around. Now for the A-4s, some of the Air Force and Marine F-4 pilots were good, but if you really wanted an expert, get an A-4 driver. After all, they were trained as attack pilots, not fighter pilots. (I don't know if it was the A-4 as a bombing platform or what, but almost to a man, they could drop snake and nape.)

I moved the team back to the western bomb crater. As they were moving, I briefed the A-4s on their target and rolled in to mark with Willie Pete. Backseat slapped me on the helmet and said, "Go! The target's cleared!" Lead was already in position to start his run. I cleared him and his napes were perfect, right next to the first bomb crater. In fact, when he released them, all I could think of was, "My God, they are going to land right on our troops!" It was that close. Both drivers dropped their nape and snakes right where we wanted them. On my BDA flyover, I saw a gook lying in the first bomb crater; needless to say, the area had been neutralized. As the A-4s were leaving the area, I happened to look to the south. Here came a fleet of Hueys with the reaction force. It looked like someone was going to get his ass in a sling. We had saved a recon team but negated the capture with our attack, and now the reaction force had been launched for nothing. However, I guess somebody felt this was not a big deal. It gave us the opportunity to put some troops into the DMZ to reconnoiter and seek out the enemy. I was on station long enough to see the force assemble and move out.

Later in the day, just before darkness set in, I was flying in an area just south of the DMZ. I observed some CH-46s going into a landing zone, presumably to pick up the reaction force. The troops were quite a distance from where they were dropped off. (I was continually amazed at how far our troops could move in a short time. They could really cover ground.)

While the choppers were on the ground, a radio transmission came in, apparently from higher up, saying the recon team was to stay in the bush. Helicopter Lead found out the team had been separated into three different choppers, stated same, and he was presently lifting out of the LZ. Higher up came back and adamantly ordered the team to stay in the zone, and if the choppers had lifted off already, fly back in and drop the team (specified by the General himself). Lead came back stating he had just left the landing zone, the situation was dangerous, and he wasn't about to take his choppers back in. There was a sneaky suspicion in the air that the recon team leader had lost his nerve, put his team on separate choppers, and was determined to stay out of the bush.

I talked to the chopper pilots in the club that night. Nobody got on their case, but as the troops were leaving the choppers after landing, a grunt Major madly swinging a swagger stick found the team leader and conducted a rather violent ass chewing. It looked like the team leader was headed straight for a court martial.

The next day, the full story came to light. The recon team point man had run into the lead element of a VC force. He turned, ran back to the team, and just froze. He wouldn't put up a fight, which is very unusual for a Marine. The guy just cowered in the bomb crater. Later when the team moved out of harm's way, the team leader had to physically drag him back to the second crater. After our attack negated any chance of a capture, the recon team joined up with the reaction force. Sometime during the day, one of the team members sat on a punji stick. The pain must have been unbearable, but he managed to stay with the bunch. When the force was being loaded aboard in the LZ, the team became separated. Hearing the radio transmissions ordering him back into the LZ, the team leader let it be known that he wasn't going back.

The enraged Major took him straight in to see the General himself. It didn't take the team leader, a Sergeant, long to convince the General that he wasn't going to take a coward and a wounded team member back into the bush. Ascertaining the situation, the General asked him if he wanted to go back into the DMZ on another mission. He did. Recognizing leadership, the General told him he could hand-pick men from any recon team in his outfit and go back in. The thing was organized that night. They were inserted the next morning and, by nightfall, had silently captured a gook—a Captain in the NVN regular Army. Yea Team!

I was in two more incidents during the ceasefire that I will never forget. The first occurred the day after we helped the recon team with nape and snakes. We received a call from a team leader located about two miles below the DMZ. He asked us if we could put him on a spot near a trail where he could set up an ambush. We found him one, and the conversation went something like this:

"Red One, go straight west from your present position. This will bring you to a large napalm burn. Right in the middle of the burn, you will see a prominent, unburned trail running north and south. Follow that trail north, about 300 yards, to a grove of bushes and trees. A good spot." (My back seat was a recon Lieutenant who knew his stuff.) The team leader acknowledged my transmission. I flew away and came back a few minutes later to find the team had walked right through the burn and were some 200 yards past the thing. I told him to turn back east and try again and flew away. When I returned the team had walked right over the trail without seeing it.

The Lieutenant in the back was making comments about the team's eyesight when I hit the mike and said, "Red One—the guy with the radio—do an **about face**" (with an emphasis on the latter). He looked up at me rather quizzically but performed the move. "Now, take fifty giant steps, **forward march**!" His stride took him back across the trail. I then said, "Red One, give me another **about face!** Now take twelve giant steps, **forward march!**" By now, the guy in the back seat was about to fall out of the bird with laughter. The team stopped right on the trail.

I queried, "Have you got the trail?"

"No Sir . . . "

I then gave the command, "LLLoooookkkkk down."

I saw his head drop and . . . "Sir, yes sir, I've got the trail!"

By then, the O-1 was shaking from the laughter. We flew away and came back a few minutes later. They had followed the trail, lost it, and were out in the open, nowhere near an ambush site.

A couple of days later, as we were flying up on the Ben Hai, I received a call from a team that was located on the bank of a southern branch of the Ben Hai. The stream was about thirty feet across and appeared to be two to three feet deep. The team leader asked me to see if I could pick up their movement across the stream. I didn't see one Marine. He called a few minutes later, stating he had some gooks on his tail and was setting up an ambush site, hoping to capture a few of his trail-ees. He asked if there was a good helo LZ nearby that he could use for evacuation if he got in trouble. It was further determined that some choppers were on the way to lift them out. There was a pretty level site about thirty yards right up the hill above his team. I saw him move his team up near the LZ. After a few tense minutes, he said the gooks had left and were nowhere to be seen.

My relief arrived on station about that time. I briefed him on the situation, and just as I was leaving the area, I saw the choppers going in to pick up the team. What happened then was told to me at the bar that night. It seems the chopper, a CH-46 escorted by a couple of Hueys, was just about to touch down in the LZ when some gook, just above the landing site, raised up out of the bush and threw a hand grenade down into the LZ. The grenade ended up right between the legs of a team member standing on the downhill side of the LZ. He simply reached down, grabbed the grenade, and "centered" it back through his legs. The thing exploded somewhere downhill in the bushes. Another team member shot and wounded the gook. Two of them ran up and grabbed him just as the back ramp on the chopper was being lowered. They dragged the guy back down the hill, threw him into the chopper, and all piled in. It was an NVN Lieutenant, carrying some highly classified documents. God bless our recon guys.

We received word on November 15 that a fellow Marine had been killed by a tiger. One must realize that the I Corps area was prime tiger hunting country. There was even a hunting lodge located in the Bai Long Valley on the Quang Tri river between Quang Tri and Vandy Base (by now, pretty much leveled by bombing attacks). The whole area was world-renowned for its tigers. It seems two Marines were sitting on the edge of their foxhole. It was a dark night, and they didn't detect the tiger's stealthy approach. Suddenly, he reached out, grabbed one of the Marines by the head or neck, and dragged him off into the dark. This was done so quickly that the other Marine didn't have time to react. Marines were immediately sent out in search of their buddy to no avail. At first light, a search party found their fellow Marine dead, at the bottom of the hill, with his internal organs eaten away. (As with most of the animal world, the tendency is to go for the stomach area first.)

There had been several tiger sightings in this particular area, so two professional tiger hunters and a Marine contingent were sent to track him down. A few days later, a recon team had set up for the night, when a tiger reached out of the dark, grabbed one of the Marines by the leg, and started dragging him down the hill. As the big cat pulled the Marine from his foxhole, his buddy tried to fire at the tiger but couldn't get a clear shot. Finally, he just reached out, stuck his rifle barrel against the tiger's head, and pulled the trigger. Chalk up one dead tiger. The other Marine was injured but survived. Minutes later, a CH-46 arrived to pick up the injured Marine, his five team members, and a dead tiger. The *Stars and Stripes* (the military newspaper) showed a picture of the tiger hanging by his hind feet. He weighed 400 pounds and measured nine feet from head to tail. You don't think that injured Marine had a war story to tell the rest of his life!

Several days later, as I was flying up Route 9, just on the edge of the Khe Sanh plateau, I spotted a tiger walking right down the middle of the dirt road. He was a big one and frightfully beautiful. I dove on him just to let him know he wasn't alone. He quickly retreated into the bush. Lord, what a magnificent specimen.

I had flown as co-pilot in the Huey a couple of times with Fritz. In combat it was strictly against regulations for the CO and the ExO to fly in the same aircraft. We took that rule with a rather cavalier attitude, and I frequently flew as his co-pilot. He let me fly the helicopter, which was an "uuuggg" to a jet pilot, but I was quickly humbled when, much to my chagrin, I couldn't land the damn thing. I could handle the rascal in the air, but as I tried to land it, about two feet off the ground everything turned to worms. Fritz would just grin, take over the controls, and land the bloody beast.

One late afternoon, after my flying day was over, Major "Dub" Shaw, our Huey Maintenance Officer said, "Lloyd, I've got a hop to take some folks from here to there. On the way back, we'll get into the landing pattern, and I'll teach you to land a helicopter." I guess he figured that if I was going to fly in choppers, I ought to be able to land the thing if the pilot got wounded. On the way back, he gave me the controls and let me fly it the rest of the hop. Flying the bird was not difficult for me, except for landing the blasted thing. I made six approaches that hop and didn't land it once! My approaches were smooth enough, but when I got down about a foot or two above the ground, the bird started going spastic. Dub had to take over each time, land the thing, and then let me takeoff and come around in the pattern again.

During my third attempt, I looked over at Dub, who was laughing like hell. I screamed at him, "Dub, this is not funny. I'm bustin' my ass trying to fly this thing, and you're sittin' over there laughing at me. Shut up!" I wasn't really mad—we were just having a little fun—but this bird was getting the best of me, and I was really concentrating. While on the next approach, I realized that both Dub and the crew chief in the back were laughing. I chastised him again. He just laughed and said, "Lloyd, why don't you take your finger off the intercom button." Hell, I didn't even know the thing had an intercom button! All my communications with Fritz had been of the "shouting in each other's ear variety." I hadn't even been briefed that such a button existed. Dub showed me where the button was: on the stick,right underneath my fingers. When we finally landed, Dub said, "All through your first approaches, you had a finger on the intercom button and were transmitting all over the helicopter. You were breathing and grunting so hard, we thought you were about to have a heart attack or a sexual climax! That's what we were laughing about." Lord, I was embarrassed.

To rub a little sand in my craw, Fritz mentioned that one of our O-1 pilots, 1st Lt. S. T. Palmason (who later would earn a Silver Star for one of his O-1 "shit sandwich" hops), liked to fly with his compadres in the Huey. I was told he could smoothly land the bird, and, in fact, made a perfect landing his first try. My attempts were now rapidly becoming the joke of the squadron, so I was determined to land the cotton picker. My main excuse was, if I landed the thing, they'd try to make a chopper pilot out of me. On subsequent hops, I watched Fritz carefully in the landing pattern. I noticed he literally froze his hand on the stick, making only very minor corrections during his landings and especially while air-taxiing the bird. I finally figured out my problem. Years ago, when my father was teaching me to drive, he told me, if I ever found myself in a tense situation while driving, to shake the steering wheel several times, left and right, ever so lightly. This action would ease my tension and make me relax. I realized I was using this little trick on all my landings—in jets as well as the O-1. Just before touchdown, I was unconsciously rotating the stick ever so slightly to ease the aircraft onto the runway. Right at the crucial second of landing in the Huey, the rotation was making me wobble all over the place. It was touchy.

One night at the Club, I told Fritz I had worked out my problem and was ready to land the Huey. He just snickered. A few days later, as we were flying back to Vandy Base to refuel, he gave me the bird and told me to land it. Man, I just froze that stick, arrived at my two feet, and smoothly let her land . . . almost. About three inches off the ground and just as I was about to relax and proclaim my first landing in a chopper, Fritz shouted, "Let's not land here—taxi it over to the fuel pits." He'd screwed up a perfect landing! I could have thrown him out of the cockpit. Recovering my senses, I made a nice transition over to the fuel pits and plunked that thing right down on the exact spot I had picked out. Yes!!! Before I left NAM, Fritz, who was a rather accomplished artist, presented me with a framed cartoon. The main theme was . . . he attempted eighteen landings and completed **one**! I'm proud of it and have it hanging on my wall.

I later returned the favor. A few weeks before leaving NAM, I sent for an advertised "authentic" WWII German infantry helmet. After it arrived from the states, we painted the squadron logo on it, had all the Majors in the squadron sign the thing, and then sequestered it away until it could be presented at a going-away party for Fritz. It was really a nice helmet. Having completed my tour in NAM, I wasn't in attendance at that party, but about one month after arriving back in the states, I talked to someone who was.

It seems Fritz was given the helmet at the club and, after the appropriate remarks, he donned it with the intention of wearing it the rest of the night. As Marine Corps parties frequently do, this one got rather wild. The tale bearer said, "Suddenly, some junior officer reached up with an empty beer bottle and hit Fritz on the top of the helmet, sending glass in every direction. Fritz dropped his monocle, didn't blink an eye, turned around to eyeball his assailant, and then turned back with a big smile on his face. I asked, "Fritz, "Didn't that hurt?" He came back with, "Didn't hurt, but it was sure loud as hell!" The bearer of good news then stated that, before the party broke up, the "in" thing was to don the helmet and have a bottle broken over your head. Many did!

One day, Fritz and I had been out almost two hours and were returning to Vandy to refuel. On our approach for landing, Fritz air-taxied behind three large CH-46 choppers. All of a sudden, our chopper started to gyrate, shake violently, and started a rapid, diving turn to the left. After the first 360-degree wild turn, I was beginning to question his motive and/or airmanship when we went immediately into a violent right turn. Red warning lights came on, beepers started going off, and when I took a look at Fritz, he was sitting there like he didn't have a care in the world! As the Huey started to right itself, I hit the mike button and said, "Gee, Fritz, that was really fun, are we going to do it again?" He just looked at me and, although he was still struggling, managed a half-assed grin. Well, we didn't crash, but we sure came close to buying a chunk of Vietnam. After the recovery and an uneventful landing, I asked him what he did to get out of the spin. He admitted, "Not one damn thing. I just froze the stick. and she came out of it by herself. God, we're lucky to get out of that one alive! Man, I should never have flown behind those big choppers. The turbulence from their rotors almost stalled us out!" Fritz made me swear before God and motherhood that I would never tell anyone about the incident, and, until now, I never have.

I was not much of a drinker, but I did frequent the O'Club every night to see the duty movie or the nightclub acts provided by our special service people. One night, the entertainers were three talented, but portly, American gals. One was more "portlier" than the other two, and the others kept joshing her about her weight. Remarks like, "you're too fat, too dumb, too ugly to get a man," were tossed about. Finally a Marine Corps Warrant Officer at the back of the crowd, who apparently couldn't stand it any longer, stood up and exclaimed, "That's OK, we'll take her, we'll take her!"

During Thanksgiving that year, although none was specified, there seemed to be a lull in the action. I hadn't flown much out to the east, so I decided to head out to the beach near the Cua Viet River and then north up the coast to the mouth of the Ben Hai. The ground between the coast and Highway 1 was level and sandy, sparsely strewn with bushes, and seemed without habitation. The terrain looked like there was absolutely no place in which to hide. I had a camera along, so I took some good pictures of the downed Freedom Bridge at the junction of the Ben Hai and Highway 1. I saw the U.S. flag flying on our side and the NVN on theirs (ours was bigger). I don't know why both sides didn't go across and blow down the flagpoles, but I guess there was a mutual pact about leaving them standing. Lord, it was some sight to see Old Glory waving in the breeze. Of course, I've always been a great admirer of the flag, and the act of burning it by the malcontents in the U.S. really stuck in my craw.

Flying south again, I caught some movement off to my left. About seven or eight troops were crouched down in the sand. They proceeded to jump up, run about twenty yards, then crouch again. They were dressed in black, so I figured they weren't ours. I called for arty, got into a discussion about being low on shells and firing limited to priority targets, etc., but I finally convinced them I had a worthy target. Meanwhile, the little people had continued their crouching and running until they disappeared into what appeared to be an underground bunker—probably a complex. I adjusted arty until I felt we had a zero and then ordered "fire for effect." One shell hit right on the top of the hole where the gooks had entered. Others tore into the sand, I'm sure caving in the sand over the domicile, and suddenly there was a large secondary explosion. A couple of shells fell a little long, but unexpectedly they caused another secondary. Lord, it was beautiful. Arty was pretty well satisfied with my BDA and wanted to continue firing. All rounds hit in the area, I'm sure caving in a few more underground facilities, but without the spectacular results of the first fire.

One day, while flying between Khe Sanh and the DMZ, I spotted a gook standing under a tree on a ridgeline about 40 feet wide between some high mountain peaks. When he finally heard my engine, he looked up and slowly moved over to and crouched in a foxhole. While I was requesting target clearance, the guy stood up in the foxhole and watched me circle. Air was unavailable, so I requested arty. After several marking rounds and corrections, we weren't even close. My target would require a direct hit. This guy's little sanctuary was in a spot where a miss was as good as a mile. I closed shop and went looking for a better target. As I left the area, I felt like hollering down to him, "You lucky bastard, this was your lucky day." About thirty minutes later, not finding another target, I returned to visit my friend. He was stretched out, sunning himself. I requested air and received two Air Force F-4s loaded with 250-pound, clean bombs. (I needed nape and snakes.)

As I rolled in for my marking run, the gook dove into his foxhole and disappeared. There had to be a network of tunnels branching off his foxhole. Screw him. I was just about to blow the whole hill down around his ears. My marking round went wide but landed on top of the ridge. Lead and Two both identified the target. I cleared them in turn, and we didn't put one bomb on top of that hill! Lead's first run dropped short, about 500 feet downhill from the target, and blew away a few trees. Two's passed about 100 feet over and above the ridgeline, then landed down in the valley, more trees. I chewed a little butt to get the pilot's attention and cleared them for another run. Lead's bombs looked beautiful but missed the top of the hill by about six inches, dropping to the waiting trees. I gave Two his correction and then watched his bombs land about thirty yards below the crest of the hill, a good solid impact, but not in the foxhole.

The F-4's runs had been pretty flat, so I gave each another crack at it, requesting that they steepen their runs considerably, hoping for a few less overs and unders. The results differed only slightly from the previous runs. We had been making the runs perpendicular to the ridgeline, so I switched the run-in heading to parallel the thing. The next two runs saw bombs dropping either left or right of the target, more trees. We had just dropped some twenty 250-pound bombs to kill one guy. My fault—I was asking for a bullseye on a target in which a miss was as good as a mile. I'm not knocking the F-4 drivers either. We had clean bombs, which were not the proper bombs for the target. As I flew over the target for BDA, I took note that the foxhole was still intact.

The next morning, I flew over the area again. The guy was standing there, washing his face out of a pot. I made a low pass on him but decided against another when he reached for a rifle. Several days later, I was in the general area when ground control called, giving me an A-4 flight that was headed home with a full load of nape and snakes, wanting to drop them instead of taking the load home. I received clearance on my ol' friend's foxhole again and headed for his hilltop. The scooters reported in, saying they were getting low on fuel and would have to make it quick. My marking round was pretty damn nice—right near the target. The flight quickly identified the target, and we dropped four napes, right on top of the hill. With little time left, we salvoed the snakes on the next run. Both birds dropped their 500-pounders right across the target. When I went in for the BDA, all I could see were bomb craters. Man, the whole hilltop was wiped out. Another waste? I can't be the judge. The little guy might have even left his hilltop, but I doubt it. His observation post had a very commanding view of a considerable amount of terrain where our troops were frequently maneuvering.

Chapter 9

Water Buffalo / Christmas

People talk about the waste of ammo in Vietnam. There was a lot, but it depended upon which side of the fence you were looking over. For instance, there were two areas on the Ben Hai that were a constant source of irritation to me. The first was located in an area where the terrain rose from the flatland to the mountains. There was a dirt road on the north side of the river that abutted the Ben Hai, about two miles into the foothills. The gooks had built up a roadbed of rocks beneath the stream's surface. When the stream was low, you could just barely make it out. When the stream was high, you would never notice the crossing. Just on our side, there existed an on-again, off-again mud hole. When this bog got too muddy to cross, the VC put tree limbs, some three to four inches thick, across the road. They had to be using this roadbed every night, and I, for one, wanted it taken out. The first time I saw the place, I called arty. The guns could be located at Dong Ha, Con Thien, Gio Lin, Cam Lo, or the Rockpile (Vandy Base). We never knew which guns would be assigned to a mission, so we flew in the general area, away from the line of fire. (On the north side of the river, after two rounds were fired, you could estimate the line of fire and stay clear.) The first round of Willie Pete landed about half a mile short near the top of a large hill. This was unusual, because arty was much more accurate than that. I gave them a correction, but the round was way over target at the base of another large hill. Another correction landed the next round back up on top of the first hill. The damn target was sitting down in a valley. Arty's trajectory just couldn't get over the top of that first hill without going over target. It would take air, which was not available at the time.

I included this little land bridge on my regular daily route, and for the next couple of days, couldn't get any air on it. Finally, one day, I came up with an Air Force F-100. I marked the target, which the pilot rogered, but, try as we might, he didn't come within 100 feet of the rock bridge. A couple of days later, I was given a flight of Marine A-4s headed back to Chu Lai, wanting to drop their bomb load. The scooter flight had the usual nape and snake bomb load, so I dropped their nape in the treeline on the north side of the river, and then concentrated on the mudhole. Lead's bombs were right on target, blowing the branch-lined bridge away. I put Two's aircraft on another target, and his bombs hit right on a nice choke point where the road narrowed between the hillside and the river. I put the rest of the ordnance on the underwater roadbed, cratering the whole area.

As the scooters left the area, you could just figure there were already some workers getting ready to fill in the mudhole, cutting limbs and branches to reinforce it, and carrying rocks out into the water to rebuild the underwater road bed, but, for a time, it would be a lot of work for them. Now, a lot of ordnance had been expended on this small target. Worthwhile? Yes. It would stop truck traffic for a day or two, meaning enemy supplies would not reach them for a while, and maybe one of our grunts wouldn't get wounded or killed because some gook didn't have enough ammo that day.

This story has been told before, but to emphasize the back-bending task the NVN performed to keep supplies rolling to South Vietnam, I'll repeat it. The NVN utilized thousands of "Chinese coolie" laborers to transport supplies to their South Vietnamese forces. It seems a little old North Vietnamese farmer volunteered his services to assist in the war effort. He reported in to his superiors in Hanoi, was given two large mortar rounds, along with the necessary food, maps, etc., to get him to his recipient mortar crew near Saigon. He took off on foot, not comprehending his life would be changed forever. For the next two months, he pounded down the Ho Chi Minh Trail, constantly hiding from air attack, bone chilled in the monsoon rains, caught malaria, had the runs the whole trip, and endured the frightening experiences of having bombs and ordnance dropped all around him during his trip. Finally, arriving at his destination, a mortar site south of Saigon, he proudly handed the two rounds to one of the mortar team members, who promptly dropped them into the tube, fired them, and then turned to the farmer and said, "Ok, go back and get us two more."

Which brings me to the second irritant. About one mile east of the previously mentioned target was an area where the Ben Hai widened and became, at high water, about four feet deep. Submerged in the river were two rope or steel cables stretching across the stream from bank to bank, one about 100 yards downstream from the other. The stream was rather swift at these two crossing points, and the ropes had been placed there to allow the gooks to hang onto something while crossing. The easternmost rope was located right where an NVN dirt road came to a dead end at the Ben Hai, so you could figure the crossing was used every night. Occasionally, a wooden raft would be tied up to one of the banks. I had tried arty and air on the blasted things but hadn't had a decent hit. Finally, I decided another tack was necessary. Heck, the gooks would just string out some more rope across the river from the bundle they had stashed there anyway, and operations would continue.

I really G-2'd the area, then met with the intelligence people to come up with a plan. I suggested a recon team be inserted on a prominent hill overlooking the cables to find out when the crossings were taking place, where the enemy were laying their caches before and after they were crossing the river, where they might be holein' up during the day to avoid detection, etc. I thought we had come up with a great plan. Well, as with most great plans, it took a while, but one afternoon, as I flew over the area, I realized it had been hit hard. The two hills on both sides of the river had numerous bomb craters. There were no ropes across the stream, and, best of all, there was evidence of secondary explosions and three truck hulls laying on the dirt road north of the river.

I found out later that, yes, a recon team had been placed in the area. The original, prominent hill was located too close to the river, and the recon team couldn't hear because of the noise from the stream. They moved to a more advantageous position and sat for three days and nights gathering information. The team expertly plotted the cables, the caches they spotted on both sides of the river, and even described hearing and seeing the blackout lights of the trucks delivering the goods. The team also determined the exact times the crossings were being made and suggested the most opportune time in which to attack. A-6s out of Chu Li carrying 21 bombs apiece were controlled by our TPQ guys (ground-based, radar-controlled bombing capability) at night, and the gooks never knew what hit them. Marine teamwork!

About ten miles northwest of Khe Sanh is a beautiful little valley—a main infiltration route from North Vietnam. It had a small creek running its length. The terrain rose gently into some nicely rolling hills, and the valley floor was a picturesque, green meadow right out of Currier and Ives. Our ground troops had been maneuvering southwest of the meadow for almost two weeks and had experienced only minor contact with the enemy. While reconnoitering over their position one day, I spotted five water buffalo lying in a mud- and water-filled bomb crater on a nearby hill east of the meadow. It was a hot day, and they were lying in about two feet of water—their own private swimming hole. I had never seen buffalo running free like this, so I looked closely for their "guards." Our G-2 people told us these animals were frequently used as beasts of burden, toting supplies long distances, and therefore should be eliminated.

I couldn't detect any movement around the buffalo, but I contacted the ground commander to see if he wanted them hit. He did. I called for air and soon had a single Air Force F-100 on station. Knowing that, if I marked the target, the buffalo would become excited and disappear into the surrounding bushes, I tried to talk the pilot onto the target. He had me in sight as I flew above the target but, for the life of me, he couldn't pick out the bomb crater. He made several passes at about 2,000 feet and then called, saying he was at a low fuel state. He pressed me to let him expend his ordnance before he left, so I dumped him on a hill a couple of miles away in Laos and disgustedly sent him on his way.

Just as he was leaving, I received a call from Major "Whitey" Meydag, a fellow squadron mate and Huey driver, saying he and his wingman had just dropped off a recon team and had nothing else to do. Whitey loved action and wanted in on the fray. I directed him over the target and then stepped aside as they started their runs. I had been constantly kidding my chopper buddies about their accuracy in delivering ordnance, so I was not really surprised when Whitey shot two rockets and both landed about twenty yards to the left of the swimming pool. His wingman's two went right. I got on the radio and chided their runs. Whitey came back laughing and said he would launch six rockets on his next run and all would be in the bomb crater. Well, I could fly just about as fast as the Hueys, so I slipped into his six o'clock position where I could get a good look at his lineup. I almost died laughing! He fired two that went left. I saw his tail swing to make a correction, and two more went right. Another tail swing, then right on target. I called, telling him he ought to trade his rockets in for a shotgun . . . might be more accurate. I dropped in behind Two with about the same results.

By now, the chaos we had created had driven all but two buffalo out of the crater. The two, obviously wounded, were straining to get up the sides of the pool to escape. I don't blame them. I was about to direct another attack to kill the wounded when I spotted a huge bull moving downhill into the green valley. I mean, this guy was a monster. He hit the valley floor, turned south into the meadow, and was loping to get away from this upheaval in his life. I pointed him out to Whitey who in turn, with great determination and eating humble pie, rolled in, saying, "I'll get that SOB this time!" He completely bracketed that bull with six rockets. So did his wingman, who went in right on Whitey's tail. Man, you can't describe the explosions, flying debris, the smoke—it was one hilarious mess. Suddenly, out of the melee came ol' bull, just loping along, head up, like nothing had happened. I couldn't believe it! He took a turn across the creek and disappeared into the bush. I never saw him again, but rest assured he had enough shrapnel in his big old body to send him to buffalo heaven! I put Whitey's flight back on the bomb crater to finish off the remaining buffaloes, even getting into the pattern to let my backseat use his rifle. At the Club that night, I told Whitey I was giving him a "Wounded in Action" for his big bull and credit to my Backseat for the last two buffalo in the pool.

I had another incident with our ground troops that again proved the accuracy of napalm and ground air teamwork. Returning to Quang Tri one day, I received a call. A platoon of grunts moving in the open, about halfway between Vandy and Quang Tri, were taking sniper fire. When I arrived on the scene, I found our troops deployed on top of a horseshoe-shaped ridgeline overlooking a small valley with its open end facing to the north (about the size of a large football stadium). The valley was completely covered with brush, but the ridgeline was completely bare. In other words, our troops were laying single file, prone, on the top row of bleachers in an open-ended football stadium and were taking fire from an area between the fifty-yard line and the goal posts. When I contacted the platoon leader, he had decided to move down into the valley but said a little air might make his task somewhat easier.

I called for air and, within five minutes, received two A-4 drivers with snake and nape, bless their hearts. I hate to give the A-4 crowd the big head, but you could almost fly over a target, spit on it, and they would immediately pick it up. This target was easy to spot, so I didn't mark it. The runs were briefed to come in from the south or right over the closed end of the stadium—and right over the heads of the troops occupying that position. Just from the tone of his voice,

I knew Lead was an old timer. Wasn't sure about Two, so I briefed Lead to drop his napes in the upper end of the bowl (the goal posts) and then Two's drop would be out in the middle of the stadium (on the 50-yard line). Lead's drop was perfect, with the napes flying about 100 feet over the top of the troops and slamming into the valley. When the troops saw the A-4 release the napalm tanks, before flying over their position, they became frightened and took off running back down the hills, away from the stadium, on a dead run. I guess it was a little disconcerting to see the first A-4 release his napes before passing over their heads. The troop commander came up on the air, screaming, "What are you trying to do, kill us?" I settled him down, and he finally agreed to let the attack proceed. The rest of the flight was just as spectacular. The troops returned to the ridgeline, and we put on a real air show for them. I gave the A-4s two KIAs, but who knows how many enemy troops were waiting down on that stadium floor? The next morning, I was told I had a ground-line call from someone from grunt land. Fearing my butt was in a sling for dropping ordnance too close to the troops, I picked up the receiver with great apprehension. To the contrary, this guy was ecstatic with praise for our work. He said he had never seen, nor realized, we could drop nape that close. They had simply walked through the area after our "surgical slice" without drawing any fire, had seen several bloody trails, and, although we probably couldn't hear it at the time, the troops gave us a rousing cheer as we departed. They had never seen a demonstration like it before. Honk! Honk!

The Marine Corps established a people-to-people program in Vietnam, assigning each squadron (or group) a local village (ville, rhymes with will). Strictly on a volunteer basis, our troops would visit the villes, dig wells, build schools, improve the local irrigation systems, and, in general, just get to know and grow with the Vietnamese. I heard many reports from our dear old TV commentators that the program was a dismal failure. Not true. While stationed at

Danang, I had a nice talk with a squadron mate who was in charge of a ville just outside Danang. TV indicated we were introducing a new strain of rice plants into the local communities but that the Vietnamese didn't like it and wouldn't eat the harvested product. The pigs we gave the farmers were of no use—the people didn't eat pork. The VC still controlled the villes and came and went as they pleased. My friend, a Captain, stated, in essence, "We have gone into our ville and dug a well that has been tested as the best in the area. The base could use this kind of water. We've built a nice one-room school-house using local materials, and the farmers have harvested their first crop of new rice. The people like it better than the old stuff, and they expect to get three crops—not the one crop they used to harvest. They are already making plans to sell the excess to their neighboring villes. Once they try pork and cook it properly, they love it. Our medical personnel visit the ville once a week. Our troops have donated enough money to buy skin diving equipment. We've taught some of the local men to skin dive, and they are going out to the China Beach area to harvest lobsters that are sold to the area's military clubs. The ville hasn't seen a VC for five months, since the popular force troops kicked them out. A few have tried to come back in but immediately get turned in by the town folks. Hey, we've got a prosperous community going, and don't let me catch some damn TV guy saying it's not working!"

One day, I heard we had a large working party going out to our ville near Quang Tri. I jumped aboard to see what was going on. Lord, the ville was within sight of the Base, but it took us a full hour to reach it. We went in jeeps, had to cross a military pontoon bridge over a large creek, traveled on what were just cow paths, and considered every black pajama-clad individual a VC!

What a beautiful little place. Our troops had done a splendid job. They soon joined in with the local carpenters to put a roof on the schoolhouse. I was introduced to the ville officials and the schoolteacher, who incidentally was a knockout! I noticed, however, that as she moved about the ville, several "older fathers" were acting as bodyguards.

I took a walking tour of the ville and found it to be right out of the Bible. Their method of irrigating the rice fields came straight out of pictures of old China. Bamboo limbs were built into a little square scaffold about seven feet tall. A young boy sat on one of the limbs, pedaled a bicycle-type contraption that was hooked to another conglomeration with large half-coconut shells dipping into the water and lifting it to a flume that emptied out into the rice paddy. Antiquated, but it sure as hell worked. Next, I saw two locals cutting a felled tree trunk with a large, two-man hand saw. That bloody tree was a least six feet in diameter and about twelve feet long. The little guys had propped it up with reed sticks at a forty-five degree angle so they could saw on it. They were cutting it into two-by-fours one huge slab at a time.

Curiosity overcame caution when I decided to travel down a dirt trail in search of adventure. I entered an open area that was covered with numerous grass-covered mounds, several over ten feet high. To the right of the mounds was a farmer driving a water buffalo hooked to a small wooden harrow. They were working a muddy, water-filled field about the size of a college basketball court. Being an old farm boy, I had to stand and watch him make a couple of rounds, fascinated by the construction of the harrow, right out of *National Geographic*. I walked through the mounds, noted the positions of several grazing water buffalo, and soon found myself at another ville, where a large group of people seemed to be conducting some kind of ceremony in what appeared to be an open-air theater.

Everyone did a double-take when they saw me standing on the fringe of the crowd. They definitely had looks of apprehension on their faces. I noticed several people crying, some on crutches, a few missing limbs—Lord, I knew I'd walked into a situation where I didn't belong. An older man, obviously a local official, left the crowd and came over to confront me. He asked what I doing in his town. After my explanation, he smiled a big tooth grin and then tried to brief me on the proceedings. It seems there was a prisoner exchange of some sort going on. Where the prisoners came from, I'll never know, but some local boys had come home from their VC prison camps and were being traded for captured VC. The local folks thought I was some high-level American official coming to stop the proceedings.

On my way back, I noticed a couple of water buffalo grazing right next to the trail. Being an old bull fighter, I stood, debating my chances of sneaking by these big creatures without being charged. The bulls hadn't really noticed me, but in a couple of minutes, a Marine PFC carrying an M-16 came up behind me. As he did, two large buffalo heads looked up. I said, "Son, I wouldn't walk past those buffalo right now. Let's wait a few minutes. They might move and give us a little more room." His answer was, "Hell, I don't care about no buffalo. If one comes after me, I'll just shoot it!" As he walked on down the trail, one of the bulls snorted and took a couple of warning steps in his direction. The PFC unslung his rifle, shouldered the weapon, but before I could tell him to lower his weapon and not shoot the animal, he turned and ran back to my position. His voice had gone up about two octaves when he said, "Sir, I believe you're right. We better wait a few minutes." We talked for a few minutes until the buffaloes grew tired of the confrontation and left.

I received a call on December 10 to recon northeast of the Rockpile and look for enemy troops. I soon spotted them resting on a hill overlooking our troops located on a hilltop to the east. There were three hills located between the two positions. I received permission to put some air on the enemy, but before it became available, I was at "bingo fuel." An OV-10 squadron mate came up on the radio as my relief. I briefed him on the mission, showing him the location of the enemy and our own troops. I cautioned him to be real careful because there were a lot of friendlies in the area. After the OV-10 driver returned to base, I queried him on the results of the hop. He said, "Lloyd, you gave me the wrong coordinates for the hill. I almost ran arty on our own troops!" I was in shock. Seems air was not available for the strike, so he called for arty. Target map coordinates were called in on every strike. Supposedly, the folks on the ground checked to verify the distance our troops were from the target as a safety factor. They hadn't caught my mistake. If air had been run, the ordnance would have been dropped in the right place, because we had both visually identified the enemy position and would have marked it for the attack aircraft. However, if arty had been run on my coordinates, it would have been on a hilltop close to our troops. Luckily, the OV-10 pilot *had* picked up my mistake. Sure humbling.

A few days later, one of our OV-10 drivers was mistakenly told he had run arty on our own troops up on the Ben Hai. Several of our troops had been hit, a couple killed. The wounded were in the Quang Tri Hospital located right on Base. He was so chagrined, he wanted to go over to the hospital to apologize and vent his grief. I told him it was up to him, but he stood a good chance of getting himself killed by some of the wounded troops, or the hospital staff, or someone else if he made the visit. He realized his position and didn't make the trip. We found out the next day from the ground commander that his troops were receiving enemy fire from across the Ben Hai River, and the friendly arty had, in fact, landed on his position. His troops, however, had been hunkered up against a bank where our arty couldn't hit them. Apparently, the killed and wounded troops had been the result of the enemy fire. He was a little pissed, though, that we hadn't located and destroyed the enemy weapons.

Arc Light missions flown by the B-52s were of no particular concern to the O-1 drivers. I had read numerous stories told by captured VC relating the terrifying experience of being under an Arc Light: bombs raining out of the sky, no advanced warning, coming from Lord knows where. We were always warned of Arc Light missions and targets on guard frequency, but transmissions could be blanked out by high terrain. You might be so busy, the broadcast might not even register in the brain, or, worst of all, we seemed to take a rather cavalier attitude about the whole thing, feeling, "Hell, I'll never be under one, so why worry about it?"

The 12th of December found my backseat and I tooling along about as far south of Co Roc as you could go before hitting the Laos border. This is some of the most beautiful country in the world, and not many people get to see it. The terrain is relatively flat coming south from Co Roc, but about ten miles below the edifice, right where the Laotian border makes a large loop turning northeast, it raises abruptly into a high, lush, green plateau, right out of an old Tarzan movie, only in color. I broke the rules one day and flew over the escarpment. Lord, it was a place you just fell in love with. There were clear streams flowing across the surface, with numerous small pools worn into the solid rock and several brooks cascading into waterfalls over the edge of the plateau. This place was so beautiful and so far removed from the war, just flying near it instilled one with a deep sense of peace and tranquility seldom known in wartime. Needless to say, during lulls in the action, I took a few flying "mini R&Rs" there to renew my juices.

Back in the real world after flying around "my plateau," we turned northeast to exit the area and were flying about 800 feet above heavily forested terrain, when an explosion went off at my ten o'clock position followed by approximately fifteen to twenty more all around us. We had just been Arc Lighted! God what a helpless feeling, sitting there with bombs bracketing the plane, not knowing what to do to get away from the thing. You could take evasive action, but what? Dive? Turn? You might turn right into the path of a bomb. I looked straight up without seeing anything and then down, as bombs kept exploding all around our little bird. Talk about two surprised and puckered folks. We were over some pretty heavy jungle when attacked. I didn't see any noteworthy targets in the area, so I just figured some bomb-laden jets were headed home and wanted to get rid of their load. Now we knew how those poor little gooks felt with bombs raining down around their ears, and, like the lucky ones, we felt overwhelmed by the fact that we were still alive.

I loved to bet with the guys who rode in the backseat with me. Most were top notch. All were officers or warrant officers with arty backgrounds, with a few coming over from recon. I really liked flying with the recon guys. God forbid, if you ever had to ditch the bird in the bush, these guys would be awful handy for escape and evasion. Few had flown, except maybe in a heli-lift in the back of a chopper, before being introduced to the O-1. The backseat windows were permanently removed from the aircraft, giving one that old open cockpit feeling that most pilots simply love. It was probably a little disconcerting to an AO on his first hop.

As mentioned, my early hops in the O-1 were in the backseat. With the windows removed, the visibility was outstanding (and it was the only seat where you could grab a rifle and start firing out at the enemy). The AOs were a great help with the "paperwork" during a flight. While you concentrated on a target, they could request the necessary target clearances, etc., from our controlling agencies to get the ball rolling, particularly when you had a hot target and were maybe already talking to the pilots, trying to steer them into the target area.

I always had a knack for sucking the AOs who flew with me into bets I couldn't lose. I mostly drank champagne while in Vietnam, so when I bet with them, the stakes were always a bottle of champagne at the club that night. After the above Arc Light experience, we flew north past Khe Sanh to the Ben Hai. As we turned east down the river reaching the flatland, I saw an alligator or crocodile swimming across the stream. I pointed it out to my AO and bet him he couldn't hit it with a smoke grenade. Surprisingly, he bounced on the wager. Now, there was no way he could hit that thing if I didn't fly right over it, so I made a watertop run, looking back to see which side he was going to pitch the grenade out. He pulled the pin, faced the right side of the bird, stuck his hand out the window, and I promptly flew over the alligator, keeping him just off the left side of the aircraft. There was no way he could come close, and he didn't. I felt a little guilty about the affair, so we split a bottle that night.

On another day, we had some troops on a hill overlooking a valley with a large, well-concealed thatched hut located south of their position, down in the valley about 1,000 meters away. We were in a free-fire area, so I started to request air. The backseat, who was in radio contact with the troop commander, informed me that the commander wanted to fire his 81-mm mortars on the hut. I had hauled parts of that mortar around while undergoing a grunt indoctrination course at Quantico a few years earlier, so I knew firsthand the difficulty of packing the damn thing over hill and dale. Although air could have easily sunk the hut, I felt those kids on the ground deserved their shot. While the mortars were being set up, I circled their position and the target several times. I bet the backseat that he couldn't hit the hut with any of the first four rounds. Being an arty type and probably manifesting a little overconfidence, he accepted the challenge.

As we circled, I asked him what his target coordinates were. He was pretty darn close. The first round landed short and to the left of the hut. He gave the mortar crews his correction: right 100, up 100. We were flying just out of range, south of the hut, so after his correction, I circled, placing him behind the mortars and line of sight to the target. Then I asked, quizzically, was he sure of his correction? He wasn't sure now from his more advantageous viewpoint. Did he want to change it? No. Then . . . yes . . . then no. The mortar crews were having some trouble with a base plate, so I flew south and lined him up with the target and the mortar position. Again, I questioned his last correction. Well . . . he wasn't sure. Then . . . yes, I'm sticking with it. I circled back to the mortar position, giving him another look. When questioned again, he stood firm, zealously saying the next round would be right on the hut. It didn't look like it would be to me, however. If the crew was having difficulties with a base plate, I figured that second round would be way off target, so I asked him if he wanted to double the bet. Sounding a little less confident, he claimed poverty until payday.

The second round was way off target. He gave the crew another correction, which I questioned. We repeated the circling scenario, but he still wouldn't double the bet. (I usually ended up buying the drinks anyway.) I kept offering corrections and joshing his, but he sat firm on his last correction. The third round was over by about 75 meters and right by about 50. He gave another correction, which I immediately challenged. The fourth round came within feet of the hut. There was some discussion on whether we had really hit the target, but, as in horseshoes, I let close count. We were close enough for "fire for effect." The mortar team plastered the area, setting the hut on fire and even getting a small secondary fire from the bombardment. My AO was ecstatic. He'd won a bet from the Major, had perfectly placed rounds on the enemy, and had enthusiastically pumped up troop morale with his glowing BDA.

That afternoon, I flew as copilot with Fritz on a recon team insert. These guys were going in . . . waaaay in . . . up on the DMZ, out near the Laotian border. We briefed, loaded up, and headed for the Ben Hai. Following the river until it became a mere stream in the mountains, we flew over a high ridge and let down into a steep-walled ravine laden with numerous caves. Up ahead, we spotted a little Army chopper (Loach), about as big as a small mosquito, that infrequently flew into our I Corps area. In fact, neither of us had ever seen one. The little guy was hovering right up against the south wall of the ravine, slowly moving from cave to cave, peeking into each as he went. We figured he was on some intelligence-gathering mission, and we honked as we went by. About two miles later, the canyon narrowed to about a freeway width, giving us both a spooky feeling. We took a real interest in the passing indentations in the cliffs, and as we passed one you could put a couple of houses in, we saw 25 to 30 people sitting there staring out at us. The area had never been entered by friendly troops, so it must have been a VC enclave. We didn't see any weapons, but we contacted the little chopper on a common frequency, warning him to be careful.

Breaking out of the ravine, we spotted our landing site, right in the middle of a large beautiful valley, a l-o-n-g way from any friendlies. I really had a "back of the neck" feeling for the recon guys. In my bones, I just knew they were being placed into strictly enemy country, possibly a VC Rest and Recreation area. However, upon checking with the recon battalion later, we found out the team had been extracted four days later . . . uneventfully. I think that day we saw a lucky Army chopper driver and a very fortunate recon team.

On the 7th of December, while flying over Helicopter Valley, named after a furious helicopter assault where we left several downed choppers whose hulks still remained as evidence, Control called saying a recon team was in dire straights. I established radio and then visual contact with the team and called for air. The team had taken some fire from a knoll just west of their position. As I rolled in for my marking run, I observed several weapons blinking up at me. For a little better accuracy, most of my marking runs pulled out at 200-300 feet above the terrain, so it felt a little hairy looking down firing gun barrels and a little more intense when placing my marking round. Definitely more determined or lucky, I put my Willie Pete right on target. The A-4s did their usual outstanding nape and snake job. When I contacted the team later in the day, they reported we had completely destroyed the area, with parts of several bodies scattered about.

About this time, Fritz and I were receiving reports from the chopper pilots about their overnight stays at Vandy Base. During the monsoon with its horrible weather, we had great difficulty getting choppers to Vandy from Quang Tri. Vandy had a nice Marston mat runway and large fueling bladders with hoses running out to fuel pits. Our choppers frequently flew into Vandy to refuel. It was also a large artillery base. Troop commanders at Vandy liked to have choppers sitting on the runway at first light to get operations under way. One story I know to be true, because I talked to the people involved, started when a grunt colonel at Vandy requested several choppers for a troop lift from Vandy to the vicinity of Cam Lo. We launched the birds, but intense ground fog at Vandy negated any attempt at landing, so they returned to Quang Tri.

This resulted in one screaming colonel calling from Vandy, saying he could hear the choppers from his position—why didn't they land? When told that fog precluded any attempt at landing, he suggested flying the birds over the runway and then dropping them vertically, straight down, through the overcast, onto the mat. When told it was impossible because the choppers had encountered a solid overcast over the area at 1,000 feet and couldn't see the ground, and the choppers simply didn't have the navigational aids to "exactly" place themselves over the runway, the Colonel made a few remarks about the courage of chopper pilots. When he cooled down, our people suggested that past weather history indicated the choppers could probably get into the area about noon, but if trucks were available at his position, the road to Cam Lo was open (a trip of about 15 miles). The colonel came back with, "For chrissakes, what are you guys talking about? The truck drivers can't even see the road to drive over from the truck compound!" Nuff said.

Anyway, to get around the weather problem, we agreed to send several choppers to Vandy about dusk to stay there the night. A Quonset hut was made available to us. We supplied it with cots, sleeping bags, etc., to make it livable, and our crews started a nightly watch at the base. They soon started to complain about the living conditions, the noise the cannon cockers made firing arty at various times during the night, etc. I finally told Fritz I would fly up with a crew to check the place out. I took a walking tour of the base to see where the real war was being fought, but when I came back to our Quonset, I found windows blown out, doors ajar, the entire place covered with dust, and generally in bad shape. I crawled into a sleeping bag and settled down for a good night's rest, thinking we were probably going to have to send a working party up to clean the place. We had enough pilots to rotate, so they would only catch the duty at the most once every ten days or so.

I went to sleep, enjoying the rest, when all of a sudden I was startled awake by the loudest thunderstorm I had ever heard. The arty big guns had opened up. Lordamighty, could they shake the ground! A guy could listen for a few minutes, get bored, and try to get back to sleep. About that time, one of the big ones would go off again, lifting you right up off your cot. I heard a few, "Oh, shits" and "Shut those damned things off!" echoing around the hut during my attempts to get back to sleep. I soon said to hell with this, somehow shut the constant roar out of my mind, and slept peacefully the rest of the night. In the morning, I found my sleeping companions rather pissed off at the whole thing. Their main comment and complaint was, "Major, we're laying here awake all night, with cotton in our ears, covers over our heads, listening to those effing cannons, and you're over there sleeping like a damn baby." Well, my report to Fritz was negative, but I felt our guys could take it. It's not in anybody's interests to keep chopper pilots up all night and expect them to be at peak performance the next day, especially when they were flying seven to eight hours every day. There wasn't any way we could renege on the commitment, but, luckily, the weather broke, allowing us the "luxury" of flying out of Quang Tri.

The O-1s seldom flew at night, the theory being "if you can't see 'em you can't hit 'em." But on the 8th of December, there was a big battle going on up between Con Thien and Helicopter Valley. We received a request to fly up and somehow assist the grunts. I grabbed the hop, but before I could launch, our returning choppers felt the firefight was over for the night. Our troops had gone on the defense, and, in fact, the ground commander had sent all supporting air home for the night. We confirmed same up through the chain of command, so I didn't get airborne. I scheduled a hop to be over the troops just before daylight to assist in any manner possible.

When I reported in to the troop commander, things were pretty quiet. He stated the VC seemed to have left the area, but he had lost two Marines during the night's firefight, couldn't get them out during the battle, and was preparing to go back in and retrieve the bodies. He asked me to look for and plot the location of the dead Marines. I found both, and it just tore my heart out. One was laying face up, stretched out on a small ridge line, completely naked except for his white skivvies and his combat boots. Oddly, his socks were missing. The other body was laying about 60 yards away, face down, half in and half out of a small stream, and also naked except for skivvies and boots. I really G2'd the battleground hoping to draw ground fire to determine any enemy present. (This was standard procedure for the Bird Dog. Many of our enemy troop sightings were after being fired upon . . . no big deal . . . but thank God and Mike Whitehouse's theory, most of the fire missed.) I couldn't detect the slightest movement in the bush, but had a queer feeling in the back of my neck that something was wrong here. I suspected some kind of ambush and transmitted my fears to the troop commander. (Everyone knows full well the lengths Marines will go through to retrieve their dead.) We agreed to run some air to shake things up.

Our troops were all located on top of a hill overlooking the little valley. They had an excellent field of vision for the impending strike and could bring their weapons quickly onto the enemy, if we scattered some. The first body was lying on a small ridge rising between two knolls, about 100 feet from the reclining figure. I planned to run the attack birds perpendicular to the ridge and drop on the most advantageous spot for an ambush—the two knolls. I soon received my favorite birds—two A-4s armed with nape and snakes. This would be a precision job. The napes would burn off the brush, hopefully to expose any foxholes, and then the snakes would tear the place apart.

Although there was no doubt the two Marines were dead, confirmed by the troop commander, I surely didn't want to put any more wounds into those inert bodies. It did, however, allow me to bring the nape and snakes in close. While briefing the A-4s, I mentioned the two dead Marines, and when they rogered my information, you could just sense their determination and heartfelt feeling for their fellow Marines. I called for two napes by Lead on his first run, just to the left of the body. Two's were to go right. I didn't even have to mark the target—that naked body was plainly visible from the air. God, what precision! The A-4s seemed to bodily hurl themselves at the target, pulling out just feet from the ground after releasing the nape. I could see several fighting holes now exposed on top of the knolls. The Marine pilots then put the snakes in, exactly bracketing the target on each side of the body. We ran the rest of the ordnance near the other body with excellent results. As I flew over the first target for a BDA, I could see three bodies lying in and on top of a trench line. Now, I have seen bodies and parts of bodies go flying through the air during bomb impacts, so I took particular cognizance of the bodies of our Marines. The ground concussion had moved the first body a little, but he didn't appear to have one hair on his body harmed. I thanked the A-4s with all my heart, ending with, "Gentlemen, two Marines are going home in a body bag, but thanks to you guys, they are going home in one piece." I received a call from the ground commander a couple of days later. His troops walked through the area unopposed and found thirteen enemy soldiers Killed By Air!

In early December, we had read in the *Stars and Stripes* that the battleship *New Jersey* had moved up into our area. I took particular interest in her because, as a kid, I had marveled at the terrain-retching impact areas of the big guns fired by our fleet onto the beaches and hills in the South Pacific during WWII. One day, after taking off from Quang Tri, I flew straight east to the beach. I could see the *New Jersey* floating offshore, and, to my surprise, I received a call from a controller asking if I had any targets for the big ship. I didn't have anything in mind nor in sight, so told him I would scout around a little to see if I could come up with something. He kept calling and pushing me; seems the ship was getting ready to leave the area and really wanted to flex some muscles. The area I was flying in was new to me and had been considered friendly for quite some time. I hustled up to the area above the Cua Viet river but still couldn't find an appropriate target for her. Control kept insisting, so I picked out a small, brush-covered hill that might possibly hold a few fighting holes. We received clearance and, giving the impact area a wide berth, I waited for the first round. This was going to be one hell of an explosion! It sure wasn't what I expected—formidable, but not like in the WWII newsreels. The first round was long. I called in my corrections and waited. Short. More adjustments. Long again. Another correction. Longer! Man, this was not the weapon you wanted on a pinpoint target.

I became a little braver as we fired and moved closer to the trajectory of the large shells. This was great. You could see the *New Jersey* fire, the smoke rising from the fired gun, then pick up the round and watch that rascal fly past your position before impacting on the target. One more correction put her fairly close, so I called "fire for effect." Man, did they make a racket. Beautiful—but there wasn't a damn thing under those bushes. I got some static from the ship. They wanted lots of KIAs and numerous bunkers destroyed, but I didn't lie on my BDAs. I had to see results before I reported them. Anyway, we closed up shop, they went their way, trying to prove their feasibility study on the capabilities of the large ship, and I went mine, knowing that "I had fired the *New Jersey*."

As I was cruising down the Da Krong Valley on December 14, I heard Fritz and one of our OV-10 drivers coordinating a recon team extract. The team had taken fire and wanted out. The OV-10 guy was out of marking rounds, so I volunteered to mark his target. He called for air. When they arrived, I marked the target and let him run the attack aircraft. Fritz's flight of choppers went into the LZ right after the ordnance drop, loaded up the troops, and headed for Quang Tri. I marked the target for another flight requested by the OV-10 driver. Another successful Marine Corps teamwork operation.

My first hop in the OV-10 came on December 15. As soon as the OV-10 contingent arrived, as squadron commander, Fritz let it be known that he was going to check out in the bird. He gave me the same privilege. Our first hop was the normal backseat affair with an OV-10 driver in the front seat. We flew our second hop with an OV-10 pilot in the rear seat. Then . . . we soloed. Fritz donned an old WWI, open-cockpit, leather helmet for the occasion and posterity. I took my bird up, wrung her out, and fell in love. Later, I found out she had one problem I couldn't live with. She was too loud—the enemy could hear you coming. The O-1 was really quiet, allowing you to sneak and poop around, frequently catching the enemy out in the open before they heard your engine and ducked for cover. I just didn't have the same luck with the OV-10, but, man, was it great to be able to shoot back.

I requested a full load of ordnance for my next hop, which the OV-10 drivers reluctantly agreed to. Heading for Khe Sanh, I was determined to "take it to the enemy." Finding no worthy targets around Khe Sanh, Lang Vei, or Co Rock, I called for clearance to hit the one remaining wall of the hotel at Khe Sanh City. I had great difficulty convincing the ground controller that this was a practice ordnance mission. Apparently, he had never heard of such a thing, but what else could I call it? Anyway he finally gave me clearance, and I proceeded to wipe the wall out of existence. Didn't hit it once—always short! No revenge. Back at Base, I cornered one of the OV-10 drivers, got briefed on the intricacies of the gun sight, and didn't have much trouble after that.

Shortly after the OV-10s arrived in country, word was passed down from the Wing: don't use the OV-10 as a close air support aircraft—restrict them to FAC missions only. It seems the only way the Marine Corps could obtain the OV-10 into our inventory was to agree to the latter. Ridiculous! Here was an armed bird we could go out in, find the enemy, and hit him before he could disappear into the bush. Frequently, before the OV-10, if you caught enemy troops in the open, before you could get arty or air on target, the gooks had dashed for the safety of the jungle. Fritz asked my opinion on the subject, me being the resident senior officer of the O-1s. I told him to hell with it, let's use them any way we want. He was of the same mind, so we gave the OV-10 drivers the freedom they deserved. Never heard another word on the subject. Hell, everyone in the Marine Corps knew we weren't going to waste this bird.

The OV-10 did take its licks when it first arrived in country. We had three get hit by small-arms fire up on the DMZ in their first two weeks aboard. Seemed natural, however: seeing a new aircraft appear on the horizon, the enemy didn't know it was a FAC bird. They had to consider the bird a gunner, so took appropriate action. Additionally, much to their credit, our OV-10 drivers were a hard-driving bunch wanting action.

One day, as I was pulling into the chocks after a mission, I heard radio transmissions proclaiming one of our OV-10s had been hit up on the Ben Hai. The pilot was wounded and was headed for Quang Tri. I listened until he landed, pulled off the runway, and stopped. I jumped out of my O-1, ran to the OV-10, opened the cockpit and shut down the aircraft. I'll never know how the pilot ever made it back. He had landed, slowed the aircraft down, and passed out. His backseater, not a pilot, had taken over the controls and taxied off the runway. The pilot had taken a round in his left knee, was still bleeding, now conscious, and suffering intense pain. Doc arrived and gave him a pain-killing shot, but as we lifted him out, the guy went through hell.

After this incident, the wing placed a "not below 1,000 feet" restriction on the OV-10 drivers, but only up around the DMZ. I guess the enemy soon accepted the OV-10 as a FAC, because they soon slacked off on the bird. (Didn't take them long to realize that firing at an OV-10 brought a bunch of steel immediately in return.)

While reconnoitering the Ben Hai one day, we spotted a ferocious thrashing around in the river. The backseat said, "Wow! Look at those two big fish fighting down there!" I replied, "Take a closer look—more like sexual intercourse to me." This brings me to the spotting of game in the I Corps area – a hunter's paradise. Besides the already mentioned tigers and water buffalo, I saw elephants, deer, wild pigs, crocodiles, and wild dogs.

One day while flying along the Ben Hai, I spotted some water buffalo swimming just offshore, and, much to my surprise, there were civilians in three boats rounding them up to get them back on dry land. I guess there are some sort of cowboys all over the world, even in Vietnam.

My first deer sighting was on December 25th, Christmas day. President Johnson had ordered another stupid, three-day "fighting holiday" over Christmas. Lord, who ever heard of taking three days off, right in the middle of a war, to celebrate Christmas? Word was passed that we could not fire back, even if we were fired upon. (We O-1 drivers were ordered never to expend ordnance above the DMZ [the Ben Hai] unless we were fired upon first. Ridiculous!) Any military man worth his salt was appalled with the bombing halts and cease-fire orders poofed out of Washington. The enemy simply took the time to regroup and move men and equipment, knowing they could do so with immunity. They never took a day off!

Mag 39 went on a "leisurely" standby status on December 22. I saw the folly of missing three days of intelligence gathering, so I told our O-1 pilots to fly if they so desired. Our enlisted maintenance people fully supported the decision. Most of the pilots flew two hops each day and came away with solid information they had come to expect but couldn't substantiate.

My hop of the 25th was on a beautiful, clear morning. As I flew up Route 1, I just couldn't believe the thousands of civilians out in the foothills, gathering wood in the area between Dong Ha, Gio Linh, Con Thien, and Cam Lo, some even up into the DMZ. As I reconnoitered up to the Ben Hai, then west toward the foothills, I saw a deer swimming toward the northern side of the river. I pointed him out to my backseat, stating, "Look at that damn traitor, he's our deer one day, but turns communist today; probably taking the lesser of two evils. He knows about the stand-down, knows we won't be firing above the DMZ, and the enemy could possibly start firing south of the river, so he's headed for the safest side." I didn't observe any "Tarzan" crocodiles creeping up on old deer, so I assumed he made it safely to Red land.

As I approached my favorite cable crossing, I dropped right down on the river. Sure enough, as I flew through a little canyon, it opened up to reveal three NVA trucks sitting out in the open, on the bank, right next to my cable crossing spot. Were they startled! About ten troops were unloading cargo into the tree line and a few more were standing around with weapons in their hands. They immediately hit the deck or ran for the trees. As I climbed for a little altitude, I circled back to find the trucks fired up, backing and turning to scamper back up north. It didn't take them long to realize I was not a threat, so they came out, waved, and ordered the trucks back. God what an insult. I gave them the finger; they gave it back. I determined I wasn't going to let the little buggers get away with their activities unmolested, so I circled their position for about an hour gathering information on their caches, the truck route headed north, just anything to let them know they'd been had and, come resumption of activities, there would be hell to pay. Having all the info needed, I searched further west for enemy activity. Finding none, and reaching bingo fuel, I flew back over my little buddies, who were sitting there resting. They had just unloaded three trucks, now on their way north, passing three more headed for the cable crossing.

The morning the cease fire ended, I was on station over the cables at daylight. I flew along just to the north of the river, trying to draw fire so I could call in air to retaliate. No soap. I knew this area like the back of my hand and had already estimated the movement across the river and possible cache sights on the south side of the river. I broke the rules, requested air, put some A-4 drivers on possible sites, got four nice secondary explosions, the two cables, and the underwater rock road about a mile up river. A good day's work, and I got the last laugh. The little buggers would have to move their caches some place else to traverse the river or at least get wet putting new cables across the rapidly flowing water.

Another day, I was flying just north of Con Thien, when I spotted a large rooster under some bushes, apparently anchored by a rope tied around his foot. I knew it was a Communist rooster because he headed for safety "just like any good VC" under like circumstances. I tightly circled his position to see if any of his friends tried to pull him into a hole. Not observing any movement, I thought about hitting the area with air or arty . . . for about two seconds. How would it sound to all of Vietnam if I went out on the radio requesting ordnance on a "known VC rooster position"? I renamed the position "a suspected VC bunker complex," ran some air on the thing, and observed a real nice secondary explosion with three bodies laying out in the open. No sign of the rooster; however, I did notice a few feathers settling back to earth after the bombing.

Don't think for one minute that the NVN didn't monitor our radio frequencies. One incident comes to mind. We were flying up around the DMZ one day when we spotted a very interesting scenario of trails converging into a large tree-covered area. We agreed on an artillery shot, and knowing that the backseat was an arty officer, I told him to handle the whole operation. Once arty is approved, the first round or, if needed, the first few rounds are Willie Peters. The white smoke is easy to spot. Backseat called for the first round . . . nothing! The radio conversations went something like this:

Blazer (Arty): Round away.

AO (backseater): Your round was not observed. Did you shoot a Willie Pete?

Blazer: That's a roger. Must have been a dud. We'll try again. Wait one . . . I've been informed we are running short of Willie Pete. Will you accept an HE (high explosive) round?

AO: OK, let 'er fly.

Blazer: HE round on the way.

AO: Sorry, but we didn't observe the impact. Let's try again.

Blazer: Roger . . . round away.

We didn't see a thing! Even an HE round is easy to spot. I called time, and we rechecked the target coordinates. The target was easily recognized, being on the junction of three small streams emptying into a nice bend on a small river. There was no mistake.

Pilot (me): Blazer, this is the pilot. We've rechecked our coordinates. I've got a funny feeling in the back of my neck that we are not on the same target. What are your target coordinates?

Blazer verified the coordinates and exactly described the target.

Pilot: Let's try one more round. Make it an air burst . . . but check and make sure your guns are pointed in the direction of the DMZ, not Danang . . . we don't want to be lobbing friendly rounds on Dong Ha or Quang Tri!

Knowing the radius of an air burst, and being a little leery about this whole situation, I took up an orbit about 300 yards south of the target. An airburst can be seen from a great distance, so I figured our chances of seeing the thing were excellent. The round was fired . . . nothing!

Pilot: Blazer, something is wrong here. We didn't observe your air burst. Let's just silently terminate this request, wrap it up, and we'll both go home and think about it. Are you sure you are in Vietnam? (That at least got a laugh.)

I still don't have an explanation. The day of the mission, the whole of I Corps were given new tactical radio frequencies, presumably to disrupt the enemy endeavor to gain intelligence by monitoring our frequencies. Although there could have been a frequency mixup of some kind, I have the sneaky suspicion our little Blazer friend was an enemy, well versed in the English language.

Before Christmas, our troops wrote home asking their folks to donate Christmas presents for our village kids. The response was overwhelming. Hundreds of gift-wrapped presents flooded our mail room. The Vietnamese didn't celebrate Christmas, but these kids were going to have a great first one. Our troops had fenced off a small playground for the kids. When I arrived, the playground was filled with bright, colorful packages. About thirty kids lined up outside the fence with that great look of anticipation on their little faces. Distribution of the presents had been well planned. Presents were placed in net sacks, each containing four or five presents, piled into age groups. A line was formed to let the kids come in five at a time, each escorted by one of our troops. Everything was in hand when the snag hit.

The Vietnamese people have a strong family, village, and caste system that we hadn't taken into account. Seems things came to a screeching halt because the ville elders wanted to discuss our plans. With an interpreter, our 1st Lieutenant in charge of the ville sat down with the respected elders to thrash things out. All the parents quietly slipped into their huts.

Meanwhile, the thirty kids had grown to at least one hundred. All our ville kids had now turned out, joined by a bunch from the neighborhood, and no local parents were available to keep them out. Many more kids were arriving, crowding the wire fence to get a look at the presents. I figured we were about to have a toy riot. Then one of our troops, getting bored with the wait, decided to demonstrate some of the toys. He picked out a skip rope, used it, then a wind-up toy. This started a frenzy in the crowd, causing a crush on the fence to run in and grab some toys. Our troops had to display anger to hold the kids at bay.

The shouting and commotion brought the elders and ville parents out of their huts. By now, facing them were at least two hundred kids straining to reach for presents. A couple of them succeeded, grabbed and run, further exciting the little people. An on-the-spot decision was made to start handing out the packages. A gate was opened to let the first batch in. Pandemonium! About fifteen kids broke through the gate, stepping on and breaking several packages, scooping up all they could carry. Our troops finally quelled the attack, and the elders went into a hut for further discussions. Their whole tack made a lot of sense: the elders wanted this to be a slow, somber, long-to-be-remembered event.

The proposal was to have the elders select proper gifts for the children, hand them out to parents, and then from heads of families down to their kids—but what to do with the influx of little people? I wasn't involved in the discussions and was getting rather annoyed watching these kids, whose number had now increased to probably 300. They were standing ten deep around the enclosure. Suddenly, out of the blue, a jeep appeared carrying a South Vietnamese Army full bird colonel, swagger stick stuck under his arm. Have you ever seen three hundred kids skedaddle? When the little buggers saw the colonel—man, it was like someone had released the flood gates. This guy looked like one tough customer. The kids all knew him and quickly decided they could do without Christmas. The elders had sent a runner to the colonel's headquarters requesting help. All it took was the colonel and his jeep driver—nobody else. (He was top dog in the whole area and had a reputation of being a "firm administrator.") The ville had been neutralized, almost in the blink of an eye. I had a hop scheduled, so I had to leave the area, but I found out later that the presents were turned over to the elders, and distribution went smooth as silk, the only disappointment being that our troops, who had worked so hard on the project, didn't have the joy of watching the kids open their presents. The Marine Corps gained a lot of admiration and love from the villagers that day.

Chapter 10

Dewey Canyon

At midnight on December 31, every rifle in the enlisted billeting area at Quang Tri started firing. Apparently, the troops had been planning this event: every shot fired was a tracer round. Man, the sky really lit up! We Majors were all sitting around the Quonset when the firing started. (Ken Buske had left the Squadron for the US of A, and we really missed his mad dash for the bunker.) The rest of us ran out the door to see what was going on. What a beautiful sight! The scene was rather sensational but resulted in one rather angry group commander. He came looking for his squadron COs, and, as our hut was the closest to his, Fritz was the first to be stormed upon. The command was, "Get your ass over there and stop your people from firing." We had been in our cups most of the evening but rapidly sobered up as we all walked over to the enlisted area. Lord, what a bunch of drunks. They sobered up as soon as they saw all the brass arriving, most of them diving into their barracks. Fritz ordered all hands out into squadron formation, preparing to give them a "dammit to hell speech," but just before he turned to face the troops, he whispered to me, "God, I hate to give these guys hell. They've sacrificed so much and have contributed so much to our success as a squadron, it's just not right to get on their backs." He didn't. He just gave a kind of half-assed speech and told them to go to bed. The whole area lit up again as soon as we left, but we sure weren't going back. Guess the group CO felt the same way, because we didn't hear another word about the incident. Sure a beautiful night!

One night, a few of us were sitting around a table in the O'Club when we were favored with a visit by a grunt recon officer. He had just returned to our area after serving a few months with a Korean ROK (Republic of Korea) outfit located just south of Quang Tri. We had just heard word that one of our chopper pilots, on his tour of duty with the grunts, had been badly wounded when the truck he was riding on hit a mine while headed up Highway 9 between Cam Lo and Vandy Base. The recon officer made the following point:

"When the ROKs first moved into the area, they experienced the same problems we are experiencing with our truck convoys going out of Dong Ha, but it didn't take them long to convince the local populace to report any VC activity in their area. We need to get tough around here like the ROKs do. If one of their trucks gets shot up or hits a mine, they simply hose down everybody in sight, including old men, women, and children. They figure some local folks knew there was going to be an ambush or saw someone dig the hole to place the mine. The locals soon learned, for their own safety, they better warn the ROKs of mines and turn in any VC they observed in the area. When I left, you could move freely about the ROK area. They were tough little fighters and a pleasure to work with . . . just like the "old Corps."

A wing directive came out in early January ordering the O-1 drivers to observe only one assigned area. I don't know who made the assignments. Luckily, my given area was up on the Ben Hai, but some of our people were given areas where no enemy activity had been spotted since Khe Sanh was abandoned. We knew our area of operation (AO) like the backs of our hands and covered the most likely areas every day. There may have been a lack of intelligence, but not due to lack of observation—rather a paucity of enemy activity. We had won the war of the DMZ. This whole thing stung: *them* assigning *us* areas. (For instance, one area covered the Quang Tri and Dong Ha Base areas, which we and a hundred other planes flew over every day.) With approval from Fritz, I gathered the O-1 guys together, briefed the situation, then told them to check their assigned areas during the first part of their hop, recheck it sometime during the hop or on the way home, and the rest of the time, they were on their own. Our guys were mighty fine pilots, had their own little suspicious areas they liked to check every hop, and became expert at finding anything unusual in the area.

One pilot, after returning to civilian life, illegally flew a little single-engine Cessna plane (similar to the O-1) under the Arc de Triomph in Paris. I saw the incident on TV, but didn't associate the name. He was later arrested for flying drugs out of South America into Florida and is presently living with the folks in the Florida prison system.

My former training as a photo pilot thoroughly prepared me for the FAC job. Photo guys use maps on most of their hops and are excellently trained in that facet. By now, I felt I knew our operating area by heart and got a chance to prove it. I was on a recon team insert, flying with Fritz. Major K.C. Carlon led the slicks carrying the troops. The insert was by way of Vandy, then south, across the Da Krong Valley, thence southwest toward the Laotian border. We encountered low cloud cover just as we approached a ridge line between two high peaks heading into the Da Krong. The cloud cover dropped as we flew toward the area. Once through the pass, we could fly to the insert point, drop off the troops, then climb up through the mess to VFR, and return to Quang Tri. No luck. The clouds dropped right down into the canyon. The flight hung around for awhile, but the weather just closed down even tighter. Fritz finally decided to give up and return to base. I piped up and told him I thought I could get us in on an alternate route. Apparently, none of the pilots in the flight had been in the area of concern. (Most of their flights, bless their hearts, were concentrated in the hot areas up near the DMZ.)

I took them west toward Co Roc, around the mountain range we couldn't get through, then back south into the insert area—CAVU the whole way—and put them right on the insert hill. When we returned to base, Carlon came up to me, rather amazed, saying "Lloyd, I can't believe you even got us into the drop area, let alone putting us exactly on the right hill. Man, if I ever need a navigator, I'll come looking for you" (or words to that effect). I looked at Fritz, winked, and as we walked away, said "Jeez Fritz, I've flown into that area a jillion times. Guess I'll have to take the brag, just don't have the heart to tell him the truth. Chopper pilots . . . I love 'em.

After we inserted the team and headed back for Co Roc, Fritz and I spotted two water buffalo grazing in some low-lying hills. Fritz could have taken the two out with his guns, but decided to let his door gunners do the job. We received clearance to hit the beasts, but for some untold reason, neither our door gunners nor the people in our wingman's aircraft could spot the two. Fritz flew over them once. The door gunners didn't fire. We swung around for another run, and, although we were at treetop level, and the buffaloes passed right under our guns—the gunners didn't see them. Our wingman didn't even see them. On the next run, Fritz briefed the crew on exactly how he intended to approach the targets, narrated our approach, but as we passed over the buffalo . . . not one shot was fired. I just couldn't believe it, but they hadn't seen one hairy part of the now-fleeing enemy. The buffalo were by then trotting along a trail. On the next pass, Fritz told the gunners to just start firing at the ground on his command. There was a lot of firing, but no results. As we passed over the buffalo, I looked back at the two gunners. They were two excited pups, firing wildly out the side doors. The buffalo still roamed. Fritz decided to eliminate the two with his guns, but by the time he swung around for another pass, the beasts had disappeared into the bush.

During debrief back at the base, we cornered the two blind gunners. They were both rather adamant, stating they never caught even a glimpse of the water buffalo. I got the feeling the two thought we were pulling their legs—that there had never been any buffalo in the first place. I said, pointing to each individual, "Remind me to never take you two deer hunting."

A few days later, I saw some gooks disappear into the bush. I called and received air, two A-4s. As I rolled in to mark the target, I spotted a couple of guns firing up at me. I shut my eyes and fired at them with a Willie Pete. The Lead A-4 driver screamed, "You're taking fire, you're taking fire! I've got the guns in sight and am at the roll-in point. Stay out of the way!" I cleared him to drop. He came in at a very steep dive angle, and I'm not sure the guys at the other end of the guns even saw him until the last second. I'm sure the gooks were celebrating, scaring hell out of one O-1 driver, then looked up and said, "OoooOOOOO ssshhitt!" as the A-4 came screaming down their throats.

While flying up around Khe Sanh (January 18), I received a call from a ground station requesting that I fly down into the A Shau Valley to conduct a recon of the area (A as in rat, Shaw as in rick). I knew this was a hot area, and we had never, to my knowledge, been cleared into the valley. It was part of the Ho Chi Minh trail system and the main infiltration route out of Laos into the Danang Hue areas of operation. I didn't have map one of the area, but, if there were enemy sightings, I could plot them on my kneeboard and orient same on a map when I got back to the squadron. Man, were there sightings! I knew the A Shau was somewhere south of Khe Sanh, so I followed the Da Krong River south. As I flew over the last ridgeline, I spotted one of the most beautiful valleys I had ever seen. This had to be the A Shau. The first thing I saw was a boat tied to a rock on the western shore of a small river, just south of the junction of two prominent but small rivers. There appeared to be no attempt to camouflage the boat, which was large enough to haul a lot of supplies or possibly ten to twelve small people. The river was so low, I felt the boat was probably useless at the time. As I proceeded further south, I noticed some communication wires mounted on telephone poles off to my left which would take looking into.

I continued down the valley, closely observing the cliffs and steep mountains on both sides of the river. I was flying at about 300 feet above the terrain. As I turned right into the first open valley area leading into Laos, I spotted some unusual road construction—never had seen anything like it before. It was a red dirt road, cleared for about 200 yards, then covered by trees and brush for another 100 yards or so, and then opened up into a road again. The road was being constructed in this leap-frog manner for several hundred yards and obviously came out of Laos.

We O-1 drivers had been briefed that the gooks frequently camouflaged their labor forces by placing foliage on their backs so, when bent over working, they looked like bushes on the ground. I was paralleling the road headed toward the Laotian border, bent the Bird Dog around to an easterly heading, and started a dive to see if the workers scattered or if I could detect any movement. There was a lot of it. As I pulled out of my dive, I received what sounded like machine gun fire from my right. I took immediate evasive action but glanced at the sound of the weapon and saw smoke coming from what looked like a machine gun nest. There were two of them located on a ridgeline just south of the road. One was located about 100 yards up the ridgeline, the other about 30 yards higher up. Now, as previously mentioned, the gooks seldom fired at a Bird Dog while we were reconnoitering. They waited until you rolled into your marking run, then opened up with everything available. Not so in the A Shau. The gooks seemed to take great pleasure in shooting at you any time they felt like it. These nests were shaped like a donut had been dropped into the ground, the hole filled up with dirt . . . and then someone ate the donut. The pedestal left in the middle was some kind of turret for mounting a weapon. (I found out later that my "nests" were some sort of AAA weapon sites, and the Valley was thoroughly covered by AAA . . . gulp! God protects all drunks and most O-1 pilots.) The gook strategy was to uncover a nest, fire, duck back into a connecting tunnel, pulling camouflage over the sight, and then hole up. I had never seen an emplacement like it before. Man, what targets! I had finally found my own personal little war and was going to conduct a full-scale assault on the enemy!

I climbed to altitude and flew back over the nests, by now completely deserted, with one being partially camouflaged by bushes.

I contacted my ground controller and requested air. After a "wait one," they came back refusing my request. Absurd! I queried the denial and was told the area was still under Army control, and we Marines didn't have permission to provide any friendly fire support. Shit. I've got two of the hottest gun positions in Vietnam, not to mention the gook road builders, and couldn't do one damn thing about it!

As I circled the workers again, they hadn't moved back onto the road but there were still stretches covered by bushes or trees. I just figured they were simply strong-willed folks, now back working on the road, unafraid of a little ol' Birddog, or, as was the custom of enemy camouflage, they had tied the tops of trees together to cover the road. (Later publications confirmed the fact the enemy would even place potted trees and bushes out on the roadbed, pulling them in during the night to allow traffic flow.)

I then flew to the south and picked up a major dirt roadway that later turned out to be Highway 922 in Laos. While over 922, I noticed numerous truck tracks, some leading into what appeared to be truck parks just off the road. I had to fly low to detect the truck parks, but I didn't receive any fire. I followed 922 to the area around Tam Boi, then north on Route 548, and up and around Tiger Mountain. I plotted the narrow road leading to the top of Tiger Mountain, noting how easy it would be to put it out of commission. I also noticed a small diesel Caterpillar tractor located about two-thirds of the way up the mountain.

It was now time to check out the previously sighted communication wires located in a valley just north of Tiger Mountain. Oddly, the poles and wires were coated with a light green substance. Looked like someone had taken a can of the stuff used to flock Christmas trees and sprayed the whole system. I picked up the communication line, completely out in the open, about halfway up the ridge line, and followed it down into the valley. It ran under the trees most of the way, but at the valley floor it completely disappeared.

With no visible connection continuing into Laos, I figured the system was old—ancient—and probably not even being used. That theory was soon brought to an abrupt halt as I turned back uphill toward the ridgeline. I received fire from what sounded like AK-47 automatic weapons. It stopped quickly, but not before I saw the smoke and plotted the position.

I decided to continue tracking the wires which disappeared a few times under the trees. After crossing a couple of valleys and hills, the blasted things dropped over the last ridgeline right down into the Hue/Phu Bi area. To this day I don't know if the wires were active, but there had been some intent to camouflage them. Later, I read in an article in the Marine Corps *Gazette* saying that the grunts found the wires and the new road under construction during the fight to take the A Shau.

I flew around the A Shau for another hour and a half plotting every piece of intelligence spotted and then headed for Quang Tri. After landing, I checked the Bird Dog for bullet holes (none . . . another thank God) and headed for the S-2 office. They were waiting, briefed me on the Marine intent to "invade the A Shau," and said they needed all the intelligence possible. Well, I sure had one of the keys. Due to my old photo pilot training, I had little trouble orienting my kneeboard plots to an appropriate map and, as I was finishing up, was told that 1st Lt. R.J. "Mo" Moriarty was about to leave for the A Shau. I gave him a brief on my hop and wished him "lots of luck." Man, did he have it! He caught several enemy vehicles out on 922, was able to get air, and bombed the hell out of them (and, much to my chagrin, the action was only about an hour after my hop.) During the interim, it seems the Army had released the A Shau to the good ol' Marine Corps. Operation Dewey Canyon had started.

We Birddoggers had strict orders not to fly across the DMZ or the borders of Vietnam unless fired upon. Mo and I had both flown rather deep into Laos and, in fact, Mo's bombed out trucks were resting in Laos! When I plotted my information on the map, I realized we were breaking the rules and might be in trouble over it, but Mo's victory probably held down the clamor. Right up until the troops were pulled out of the A Shau, we Birddoggers flew into Laos without sanctions . . . but hell, air and arty continually flew and fired across the border. (I hope we never see another war where politicians stymie military aggressiveness by instigating the ridiculous "Rules Of Engagement" clasped on the military in Vietnam.)

I flew back to the A Shau (January 19) with great expectations of getting even. I went for the communication wire area first and put a flight of Scooters with snake and nape on the area I had taken fire from the day before. No one in the flight took enemy fire, and we really cleaned the trees out. As occurred too frequently, however, I couldn't find any bodies, spot any weapons, or see any sign of the enemy. The gooks just seemed to go to ground during air and arty attacks and then emerge after the dust cleared. I put another flight on a couple of ridges where the wires crossed, knocking down the wires and poles at what I hoped were the toughest hills to climb to repair the wires. Time to hit the machine gun nests and the new road. Another Bird Dog had run air on the two machine gun nests, completely obliterating both, so I tried my air on the workers. Many of the "bushes" were not out on the road, but I selected some real nice choke points, nicely cutting the roadbed.

A Shau again! We Bird Dogs didn't forget other areas in the northern I Corps area. We usually flew two or three hops a day and usually made one to the A Shau. There was a real fight to see who got to go there—everyone wanted a piece of the action. I expected to run air on the worker area again, but none was available. I flew northwest from the road up around the Co Ka Leuye mountain range and then to an area just north of the old machine gun nests. I spotted some trails leading east from the foothills near Co Ka Leuye. They crossed a large valley and then disappeared into the triple canopy. It was pretty obvious they continued up to a huge granite outcrop sitting on a mountain peak overlooking the entire Da Krong Valley. I then reconned 922, Tiger Mountain, and my own old haunt, the comm wires. The comm wires we'd hit the day before were still down and no enemy fire. The gooks were learning: shoot at a Bird Dog and you were soon going to have metal raining down from above.

Before takeoff on January 21, I received map coordinates to run air on top of a ridgeline in the A Shau and blow the top off, specifically to create an LZ (landing zone). The LZ was the future home of ARVN (Army of the Republic of South Vietnam, the friendlies) troops supporting the Dewey Canyon Operation. Air was on station when I arrived, and we started building the LZ with 1,000-pound bombs. The only pressure involved with the construction of an LZ from the Bird Dog viewpoint was . . . it better be in the right place! As far as I know, we were never wrong. While in my marking run, I received a call from an OV-10 driver saying I was taking fire. He had spotted the area it was coming from, had rolled in and taken the gook weapon under fire. We coordinated tactics. When I or one of the attacking aircraft were in their run, the OV-10 would roll in and strafe the gook weapon. It worked perfectly, and, after my run, we never heard from the gun again.

After doing my part, I headed for the workers. Another Bird Dog had really hit that area and 922, and I couldn't really come up with a worthy target. Usually, in this case, you went back to an area you had previously worked: the new road. I followed it into Laos, spotted some twenty or twenty-five troops walking down the road, but I couldn't get clearance to hit them. These guys didn't even head for the bush as we flew over. Being out of rockets, we threw a smoke grenade at them just to let them know we were around.

The ARVN LZ was designated FSB Lightning, and the next day I was assigned the task of controlling the insert of the ARVN troops. While the helicopters circled, I ran air all around the top of the ridgeline in case there were gooks waiting to hose them down as they attempted to land. The birds then went in unopposed. A perfect insert, except later I was told by the Huey Flight Leader, Capt J. R. Mires, that, after being told to take a couple of week's supplies, the ARVN had only brought along a couple of days of food and water. I can't tell you how this must have upset the powers to be and especially the chopper pilots who had to continually risk their butts flying back into the area on resupply missions. We sure could have used the choppers for other missions.

My first hop into the A Shau in an OV-10 was January 26. Arriving in my old road building area, I received a call from a couple of Hueys who were taking fire from the southern base of the mountain. I figured they were on a recon insert and had been shot out of their landing site. The chopper guy wanted me to fly cover for them and, if they received fire, roll in and take the area under fire. I got my rockets armed. When the Huey reported ground fire from his three o'clock position, I spotted smoke coming out of a tree line. By now I was fairly accurate with the OV-10 weapons and plastered the area with six rockets, screaming "Gotcha" all the way down during the run. I'll never know if I hit anything, but the firing stopped. In the OV-10, you had the opportunity to fly over your target and access your own BDA. Only you and God would know the results, but rather than report 100 enemy killed and numerous bunkers destroyed, I didn't even report it. The smoke was in Laos anyway. Later in the Dewey Canyon Operation, our ground troops were given D.C. permission to cross 922 and the Laotian border. (*Don't Tell America,* written by Mike Conroy, is a must for anyone connected with the operation and is an outstanding account of Operation Dewey Canyon.)

Straight north, a few miles above Khe Sanh, the Ben Hai narrows into its various headwaters. Much to my surprise, while reconnoitering the area, I saw huge paths meandering through the tall buffalo grass. This was something new and had to be investigated. I made several passes at ground level, no firing. Then I spotted the culprits: two large elephants standing near the tree line, stripping leaves from trees with their trunks. I guess the insect problem was running a little high in their paradise because they were covered with the red dust prominent along the DMZ. They frequently dusted off their huge backs with red dirt to keep insects at bay. If the large ones hadn't moved as I flew over them, I might have never seen them. I even had difficulty pointing out their location to my backseat. We circled the area for another twenty minutes. They paid us no heed. I hadn't spotted any other movement around them, so I left them in utter content. Back at the squadron, I encountered my good friend, Warrant Officer Luckenbaugh, who had flown with me more than any other AO. I didn't give it away but gave him the coordinates of the elephant herd, telling him he might find something of interest in that area.

When he returned, he was all excited. In the S-2 office, he proclaimed he had found a major VC R&R area, and there had to be several enemy platoons hidden in the little valley! I let him go on for a few more minutes and then said, "Did you see the elephants?" It took a few seconds to sink in. He said, "Elephants?" I replied, "Yes, I sent you up there to take a gander at two elephants, roaming all over that valley in the buffalo grass, and I thought you might see them and be able to say for posterity you had seen elephants in Vietnam!" As he rethought his sightings, he realized they were elephant trails. Then both he and his pilot got upset because we had seen the two large ones and they hadn't.

When you sighted beasts of burden like these elephants, you had to assume there were VC in the vicinity, and although not garnering any other sightings to substantiate enemy activity, it certainly couldn't be ignored. Knowing that arty couldn't reach down into "elephant valley," we suggested that S-2 request a high-altitude TPQ drop that night, not specifically for the elephants, but for any enemy troops that might be in the area. All was not for naught, however: the AO's Birddogger took him back to the area the next day, and they saw the elephants and no other activity. I didn't fly into the area for a couple of days, but, when the opportunity came, I saw no signs of our elephants or bombing suggestion. Those two elephants are probably still alive today, if they weren't worked to death before Saigon became Ho Chi Minh City.

One day as I was flying into the A Shau, I received a call from a ground station requesting me to spot and call in arty. "Any target would do," which was very, very, unusual. Sometimes it was like pulling teeth to even get arty to fire for you. I called in the coordinates of the worker area, received target clearance, briefed the arty folks on the target specifics, and called for a marking round. I received four rather than the usual one! I called to verify the number of rounds fired. The answer was still four, and I figured they knew what they were doing. (I suspected it might be an Army unit about to leave, and maybe they had a lot of arty rounds they didn't want to lug home. I had heard the Army did a lot of overkill on their targets and had a "get rid of 'em, don't pack 'em home" philosophy.) Whoever it was, we sure blasted hell out of a lot of trees and possibly a few gooks. My BDA was rather lacking in grandeur, discouraging to the cannon cockers, but again it was just impossible to see down under the triple canopy. I was told I had been firing two batteries. Lord, what a luxury, or maybe a waste . . . who knows.

By the 28th of January, our troops had opened up FSB Razor and Cunningham. The easiest way to tell which direction the Razor Cunningham arty was firing was to fly up the Da Krong Valley, below and abeam of the two FSBs, and see which way the guns were pointed. If they were firing, it was simple to spot the impact area and stay out of or under the line of fire. A ground station called, telling me we had Marines "in the assault" located on the main ridgeline between two peaks across the valley and west of Razor. They had taken fire, and they wanted some air. It arrived in a big way. I could easily pick out the troop lead elements, ran two flights of scooters with snake and nape just in front of the troops and up the ridgeline, top, and both sides, ending up "and then some" the other side of the objective. The scooters were magnificent, as usual. When I exited the A Shau about two hours later, the ground station thanked me and said our troops had taken the objective without a shot being fired. Oh what a feeling.

I had a great day flying with Fritz—six long hours of Huey (January 29). We flew cover for a recon insert . . . successful; cover for a supply hop into our troops located up in the DMZ . . . not so successful; and the CH-46s got shot out of the drop zone by enemy mortar fire coming from north of the DMZ. We escorted a Huey slick loaded with a recon team that had been extracted in a hurry the night before but left their radio in the LZ . . . partially successful, as we made firing runs around the LZ before the team was inserted. We took ground fire. Fritz and our wingman really hosed down the area, the team was dropped in, miraculously they retrieved the radio, and everyone got out safe. On shutdown back at Quang Tri, we discovered one bullet hole in our Huey, back in the tail section. I guess you could call that one successful, too.

The evening of December 31, just before dark, our flight surgeon, Navy Lt. "Doc" R.A. Ferguson, came into the Majors' hut, sat dejectedly on the bed, and asked to see the CO. I found Fritz and brought him back to our corner of the Quonset. Doc had a story to tell, but he didn't want it to get out to the rest of the squadron yet. Seems we had lost a Huey right into a fire fight a couple of hours ago. Some ARVN troops led by two Green Berets were in the vicinity and had driven some VC out of the crash site. Their report was . . . no survivors. CH-46s had gone in to retrieve the bodies of our chopper crew, had brought them all out in body bags, and dropped them at Doc's medical facility. When he opened the bags, all he had was one charred body and three decapitated heads. He didn't know what happened at the site, but he felt the VC had attacked the downed chopper and made sure the crew was dead, possibly by decapitation.

Our first reaction was . . . grab your rifles, let's jump in a Huey and go get the dirty bastards!

Fritz decided to tell the other Majors in the Quonset, whose reactions were the same as ours, but Fritz got everybody calmed down to await further confirmation. It came. An ARVN VC battle was raging below the chopper when it was hit. The Green Berets saw the chopper go down . . . and so did the VC. The ARVN had driven the VC, who had set up an ambush around the chopper, to higher ground. When the two Green Berets arrived at the chopper, all they saw was devastation. All four crew members were burned beyond recognition. As they tried to pull the pilots out of the wreckage, flesh and bone simply came loose and clung to their hands. Just about then, the VC staged a counterattack. Being in an untenable position, the Green Berets did the only thing they could do to help identify the bodies: they sent out the heads, teeth intact, and headed back into the bush. (Dental records were the primary means for identifying individual bodies.) When the CH-46s and their Huey escorts arrived at the scene of battle, they drove the VC back into the hills. The body bags were loaded into a CH-46 by the Green Berets, who then went back into the fight. Later, identifying the gravity of the situation, we dissolved our "revenge attack" on the VC.

I couldn't fault the Green Berets' decision under the circumstances. While at Camp Pendleton, I stopped to look at a helicopter crash, about thirty minutes after it happened. The chopper, with six Marines aboard, flew into some high tension wires strung across the Camp Pulgas road. The aircraft got wrapped up in the wires, broke off a rotor blade, burst into flame, and came crashing to the ground. The bodies had been badly burned, the flesh just melting on their frames.

The above might be a little gruesome to the uninitiated, but one of our officers who had been on recruiting duty in a large metropolitan city told us about a few incidents he'd encountered. Part of his duty entailed informing families that their sons had been killed in action. He said the local people soon recognized his visits, some falling down on the floor, writhing in sorrow, others physically attacking him as he approached the house and, worst of all, telling mothers and fathers that coffins couldn't be opened.

During the viewing before one funeral, the mother wanted the coffin opened so she could see her son for the last time. The officer refused. She became incensed, threatened lawsuits, then physically attacked him. While others held him off, the funeral director opened the casket. When the mother peered into the coffin, she screamed and fainted dead away. The body had been savagely burned, and all that remained was a charred torso and head, the limbs having been burned into stumps. Everyone at the funeral would live with that experience the rest of their lives. Another officer I talked to had been assigned to escort the body of a dead Marine to his home for burial. Not having seen the body or being familiar with the situation, he relented when the mother pitiously insisted on opening the casket. They did. It contained a burnt head and arm . . . the only remnants of this fine fighting man.

Fritz's baldness became a grave bone of contention that night. We had held a solemn drinking wake for the dead chopper crew and all went to bed early. Sometime during the night, a mouse fell from the rafters of the Quonset hut onto his bald head. Fritz came out of his dreams slapping to get whatever it was off his "revered spot." The poor little mouse was running for his dear life but couldn't get traction on the slick surface . . . just sat there spinning his wheels. The commotion woke the rest of us, and Fritz proceeded to tell each in turn his story. The more he told the story, the larger the mouse . . . it ended up being a large rat! Fritz was rather unnerved about the experience and vowed that it would never happen again. He solved the problem by asking his wife to go out and buy the thickest stocking cap she could find. After it arrived, Fritz put it on every time he walked into the Quonset, his theory being, if another rat came flying out of the ceiling onto his head, the stocking cap was going to provide him with a nonskid surface to run on!

During the week of February 3 - 8, the A Shau really socked in. The low clouds enveloped most of our FSBs in the valley, but, if luck held, we Bird Dogs could usually fly up the Da Krong, right down on the river, pass below FSB Razor, then drop down into the valley and go looking for targets. (The weather was usually much better around Tiger Mountain and Highway 922.) Not so for the choppers. Even if they found a hole to crawl through, the FSBs were still socked in, and the goods couldn't be delivered. They still tried. (Just prior to Dewey Canyon, there had been rumors circulating amongst the chopper guys to the effect that Major General R. G. Davis, CG 3rd Marine Division, was down on chopper pilots, thought they didn't have the guts to fly tough missions, wouldn't go into hot LZs, etc. In my opinion, the General had been using the hell out of choppers. I told them so and added, "If the General didn't like you guys, why would he base the entire Dewey Canyon operation on the use of helicopters?"

The gooks soon caught on to our little Da Krong route, and choppers started taking fire from the sides of the canyon. Finally a chopper was shot down, prompting the powers that be to promulgate the edict that there would be no more low-altitude flying up the Da Krong. (It didn't stop the O-1s. We figured they were only talking about choppers.) That night I was sitting with a bunch of chopper pilots discussing the restriction. Their general consensus was "F" the Generals! There's only one way to get into the A Shau when it's socked in, and that's up the Da Krong. "We're going up the Da Krong!" Chopper pilots . . . the greatest!

One of the worst decisions I made in my entire life occurred during the period of February 10 - 20. I came up for another R&R. My own little war was advancing splendidly, but my "spunk level," as the chopper pilots called it, was extremely high. War or no war, I needed to get back to Australia and its lovely ladies. After picking up the necessary papers, I fired up a sick O-1 for delivery to the maintenance people at Monkey Mountain (Marble Mountain) located near Danang. The hop down was wonderful. I took the tourist route down Highway 1, basking in the almost relaxed view of the terrain. Still, once a Bird Dog, always a Bird Dog, and I found myself automatically looking for targets. (Even today, as I drive down the road, I find myself looking for targets, especially when traveling through mountainous terrain.) After landing, I caught a bus over to the passenger terminal at Danang and found out there wouldn't be a hop to Australia for two days.

Having never been to China Beach, I decided to spend the day basking in the sun and visiting the big PX. The China Beach complex was located several miles east of Danang, reached by going through the outskirts of Danang City—Indian country, as far as I was concerned. I caught a military bus driven by a Vietnamese civilian and jam-packed with Marines. Almost the last man aboard, I lucked into an aisle seat, front row right. As I settled into the seat, I noticed a Korean ROK (Republic of Korea) soldier standing up next to the driver. I had never seen any ROK troops before, but they had a rather fierce reputation for shooting first and asking questions later. I enjoyed the ride but had thoughts in the back of my mind about how easily it would be for a VC to zap a bomb under the bus and blow us all to kingdom come.

We soon crossed a river on a military-constructed bridge and stopped at what was presumably a bus stop. Looking out the window to our right was an ARVN (Army of the Republic of South Vietnam, the friendlies) outpost with several troops lounging around going about their daily business. The bus was filled to capacity, so when two ROK troops jumped aboard, they joined their buddy standing up front. They all had .45s strapped to their hips, the only weapons aboard the bus. As the driver pulled out into traffic, we heard rifle fire, and out to the side about twenty yards, several ARVN were running in the direction of the bus firing their weapons into the air. Someone in back hollered,

"They're shooting at us!" I would have to guess that all the Marines and the three ROKs now standing in front of me were seasoned combat troops. They hit the deck. Ol' dumb me, who had only been shot at while flying an airplane, looked around to see what was going on and saw nothing but Marines hitting the deck. I finally piled all 210 pounds of my raw flesh down on top of a couple of Marines who were lying on the floor in the aisle.

As the bus slowed, the firing slackened, so I slid back up into my seat only to find an ARVN soldier demanding that the driver stop and open the door. That accomplished, the ARVN jammed his M-16 in the door and started jabbering. I certainly couldn't tell what was being said, but the ROK trooper closest to the door suddenly un-holstered his .45, pulled the slide back, and pointed it at the ARVN. Man, the shooting was about to start, and I was one of the biggest targets around. I knew something had to be done, so I jumped in between the ROK and the ARVN. (Ever hear the one about being between a ROK and a hard place?) Facing the ARVN, I moved down the steps toward him, saying, "Put your gun down, put your gun down," with all the conviction I could muster. He didn't exactly know what I was saying, but I kept motioning him to lower his rifle. Guess I looked like a giant to the ARVN. Probably figured if he shot me, I might fall on him and break his neck, so he lowered the weapon, slowly, from my face down to belly high, then finally to the ground.

As we were standing there trying to sort things out, a young Marine stepped out of the bus door. I said, "Son, I believe you better stay aboard the bus." He answered, "Oh, I'm just getting off at the next bus stop, and I can walk from here." As he turned left toward the front of the bus and walked about four feet, another ARVN soldier stepped in front of him and stuck his rifle barrel right up in the young Marine's face. The kid turned back to me and said, "Well, I guess I'd better get back on the bus!"

The original ARVN kept trying to get around me and back into the doorway, but I planted my big bod in his path and wasn't about to let him by. Presently, a uniformed American warrant officer, maybe a SeeBee, came on the scene, asking me what the trouble was. Well, I had no inkling of the situation but told him our side of the story. His appearance was somewhat rumpled, and he smelled to high heaven of booze. The thought crossed my mind that here was a guy who was going to be no help at all. Surprisingly, he turned to the ARVN and started sputtering what appeared to be fluent Vietnamese. The ARVNs still didn't seem to be appeased, rifles at the ready. Suddenly four or five heavily armed Marines appeared out of nowhere and greatly relaxed the situation. One of them spoke Vietnamese and questioned the ARVN who seemed by now to be the leader of "the enemy." The ARVN troops relaxed their warlike stance as the peace negotiation went on. The welcomed Marines were a team who lived in the area—our people-to-people program.

In essence, the ARVNs claimed that, when the bus stopped and just before they boarded, the Koreans had turned and fired into the ARVN outpost. This was ridiculous, because the ROKs had walked calmly onto the bus, guns holstered, and exchanged greetings with the Korean already on the bus. Finally, the negotiation seemed to be in our favor, and the sergeant in charge of the armed Marines told us we were free to go. (If you are ever in a jam, call a Marine Corps sergeant!) The driver shut the door and proceeded out onto the road again. As he did, we heard firing from the ARVNs again! As everyone hit the deck, we screamed at the bus driver to "Hy yok o, Man" or "keep going, get the hell out of here!" He did, and we soon put some distance between our bus and the ARVNs. As I looked back into the bus, I spied the three Koreans sitting on a seat about halfway back on the bus, on the opposite side from the firing. I was told that when I stepped between the two combatants, the Marines aboard the bus had literally lifted the Koreans further back into the bus to protect them from the ARVNs. When I saw them huddled down in the seat, I said, "Were you guys scared?" Their answer was a blank quizzical stare. I then took off my hat, pointed at them, motioned that old American habit of wiping the sweat from the brow, and said, "Sshheewoooo." They got a big smile on their faces and gave some happy nods, signifying yes, they understood!

What a fiasco! I had been shot at many times in the O-1 and probably came close to buying the farm, risking my neck almost daily supporting our troops, and here I had just escaped a shot in the gut or the back on a goddamn bus headed for the China Beach PX.

When I arrived at the terminal the next day for pre-booking for my hop to Australia, there must have been a thousand troops waiting. Two birds were scheduled to take us south to the land of *sex*. Only trouble, there were too many troops waiting with bated breath. I had already been to Australia, so I felt a little awkward about being given "officer" priority over the troops standing in line, and, like an idiot, I held back. Finally, it came to four seats on the last bird. There were five of us left. I was first in line, but as I looked around at the others waiting, they were in beat-up utilities, covered with the red dust of Vietnam, and they had to be front-line troops going on their first R&R. These guys certainly deserved Australia more than I did, so I told the boarding clerk to let one of them go in my place. Funny what the ol' conscience will do to you sometimes.

I hung around the terminal to see if everyone showed up for the flights, and they did; however, I learned there was a bird leaving for Kuala Lampur, Malaysia, in about an hour. I thought about getting back into my war, and should have, but then decided to see what Malaysia had to offer.

After landing at the airport, we were whisked away to motels that were firmly entrenched in the red light district of town. Each had its own little bar and dance floor and some rather beautiful bar girls. I just couldn't see getting into that scene, being more inclined to the "round eye" types, so I decided to go to the best hotel in town for my seven-day stay. The place was a great and comfortable hotel with an excellent rooftop restaurant. From the look of things, though, the hotel was seldom over half full.

For the first two nights, I was the only guy in the rooftop restaurant at dinner time. An ol' VMCJ-er, Lt. Shipman, an R/O on the same flight as mine, had decided to stay on the raw flesh side of town, but the third night he appeared in the hotel. We had a couple of drinks together. I briefed him on the niceties of the hotel. He went back for his toteables and checked into my hotel. Later that night as I entered the restaurant, he was sitting with a priest whose denomination has escaped me, but what an interesting gentleman. He had been posted in Africa for around thirty years and had some very interesting stories to tell. I got in a few rodeo and airplane stories. It was a nice night.

I soon found out there was a race track on the outskirts of town. Being an old horse racing aficionado, the next couple of days were spent at the track, where I found some rather pleasant ladies.

The funniest experience during my stay was when I tried to call home to the US of A. My operator spoke excellent English, but there were a few phrases she hadn't quite mastered. My call was to Utah. She politely accepted the number, but I had a hard time convincing her there was a Utah in America. We finally got it all worked out, but she said it would take about thirty minutes to place the call, and she would call me back. A while later, she called, saying your call to "ooo tah" is ready, sir. I took her to dinner that night.

I was always prowling the hotel for a warm body, but I'd seen very few women on the premises. One night at dinner, however, the restaurant started to slowly fill with what appeared to be Japanese businessmen. They were soon followed by natives and girls of all shapes and sizes. The little hostess seated each girl with her "fellow," and in about an hour, the place was literally overflowing. Those poor Marines out in the district had lost their "honeys" for the night. It turned out to be a swell party.

One of the highlights of the R&R was the celebration of the Chinese New Year. I happened to walk out of the hotel one night and saw what looked like millions of firecrackers hanging in streamers from poles draped out from the windows and rooftops of the tall buildings in the business district. Employees from the hotel were placing long streamers of fireworks across the steps leading into the hotel and hanging others all over the facings on the entryway. I soon saw them place huge streamers coming down from the roof. Thinking this might be fun to watch, I waited on the sidewalk. At the appointed time, the fireworks were lit. I was astonished and had never seen anything like it! Man, when they lit the streamers of firecrackers, it was positively spectacular! The explosions started at street level, worked upwards in an exquisite display of noise and fire, most taking five to ten minutes to reach the top, which contained the final "coup de grass"—a huge ball of firecrackers blasting off all at the same time. Utterly fantastic! Kuala Lumpur was a nice, pleasant interlude, but boredom set in after five days. I was itching to get back into the war. I should have stayed in Quang Tri for Dewey Canyon.

Back at Quang Tri just after night set in (February 21), we started receiving rocket fire from across the river. Major Tod "Ike" Eikenbery had the nightly helicopter alert duty. As the shells were landing, Ike was launched, directed downwind, and started firing runs on the VC launchers. I wasn't at the O'Club during the melee but was told later that, instead of running for safety, most of the officers sauntered out of their hooches and the club, stood on the edge of the river, and each time Ike rolled in on the target, saluted him with raised glasses and gave him a rousing cheer. That's pilots for you. The next night, the base received three enemy mortar rounds . . . not enough to get out of bed for.

My first hop after arriving back in Quang Tri was back into the good ol' A Shau. Things had really changed. Our troops were running all over the mountains and valleys. Thanks to our Birddoggers, air, and arty, Highway 922 had several disabled vehicles strewn along its thoroughfare, and the bulldozer on the road leading to Tiger Mountain was laying about fifty feet down the south slope. (I had missed a lot of action.)

It took me about an hour to locate and plot all the friendly positions. While doing so, I spotted another one of my machine gun nests on a hill in a valley northwest of Tiger Mountain. I called for air. While waiting, I saw two choppers flying north up the valley. Suddenly, I heard firing from the area of the nest. The choppers must be taking fire. I swung around and caught a glimpse of smoke remnants rising from the gun that fired. I quickly broadcast a warning to the choppers, who had by now proceeded north and were out of gun range, telling everyone who would listen that I had an active gun position just north of Tiger Mountain. My AO called for arty. Nothing was available. About fifteen minutes later, while we were waiting for clearance to hit the area, two more choppers came along. I tried to inform them of the danger on guard frequency, but they didn't appear to have received my transmission. Alarmed at their failure to take evasive action, I looked down and saw camouflage being pulled off the gun pit. There was only one thing to do. I armed both firing switches, told the AO to hang on because we were rolling in on the gun, and proceeded to dive at the nest. On the way down, I fired all of my rockets and got rather low. Guess I was so focused on the dive, I didn't even notice if there was a weapon in the nest. I do know that the weapon never fired during or after my run . . . so at least the choppers had a safe trip.

We never did get air or arty on that little pit. I kept a weary eye on it until it was time to leave. On the way home, I contacted another inbound Bird Dog, briefed him on my "nest," the danger to the choppers, and the requests in for air and arty. After shutting down the O-1 at Quang Tri, the plane captain came around to the cockpit and said, "Sir, you've got mud all over the engine nacelle, and there is some white stuff in the mud." Sure enough, there was mud on the little bird. I told him the white stuff was from a Willie Pete . . . not to touch it . . . asked him to quickly get someone to hose down the little bird, and swore him to secrecy. I didn't want to get grounded for stupidity. I must say, my AO turned a little white when he saw the mud on the aircraft. The next morning, I headed right back to the nest. It was still there—seemed inactive, no choppers were flying up the valley—but I put in a request to hit it anyway. Air took it out beautifully. Scratch one "nest" and add a nice secondary explosion. Somehow my ground controller had observed my dive on the nest the previous day. When I gave him my BDA from the air strike, he asked, "Were you the guy who made that run on the target yesterday?" I answered in the affirmative. He said, "Man, when you make a run, you really make a run. It looked like you had crashed!"

While headed up Highway 9 one day, between Vandy and Khe Sanh, I spotted 31 folks walking down the road toward Vandy. They looked like locals but were in a free fire area. (The Northern portion of I Corps was designated a free fire area. All the civilians had been pulled out, and we were cleared to hit anything that moved on the ground.) I called in the situation. Trucks were sent out of Vandy to pull them in. I found out later they were Montagnard people who had been chased out of the mountains by NVN troops.

February 28 was a busy day. As I flew into the A Shau on a beautiful clear day, we received a call to assist a recon team ("Snub Nose") that was taking fire. I put in a call for air and five minutes later was over the team. They were located on the previously mentioned granite outcrop east of Co Ka Leye. The outcrop was about the size of a western barn and a couple of outhouses. On the eastern edge of the outcrop, a ledge dropped off about four feet onto a small rock platform, maybe 50 feet in width. From there, the terrain took a sheer drop, 700-800 feet straight down, right out of an old Tarzan movie—remember the escarpment? Anyway, for any Eastern boy who had never seen a tall mountain or stood looking down into the canyons of Yellowstone or the Grand Canyon, it must have been awesome, especially to the Snub Nose team. The entire western side of the mountain was covered by dense jungle. There were several well-worn trails leading up the side of the mountain that disappeared as the foliage grew thicker. All, however, seemed to lead and converge at the granite outcrop.

Snub Nose was located up on that granite knob, and when I raised him on the radio, he said they were taking fire and about to be overrun. When I asked him where the firing was coming from, he said, "I'll show you!" He threw a red smoke grenade as far as he could toward Laos and said, "They are right where the grenade landed." This was serious! I screamed to expedite my air. It would be there in ten minutes . . . but I needed it now. I made a run on the position but didn't hear any firing. I rolled in again, thinking my activity might keep the enemy from advancing. I suddenly saw and contacted two Huey gunships flying up the valley and requested their assistance. Just before they arrived, I received a call from my air—two Air Force F-4s. What I really wanted were some scooter drivers, because these drops were going to be close in.

Now, it's common knowledge among FACs, when you talk to another pilot in a "shit sandwich" situation, if he comes up on the radio with a smooth, effortless, non-stressful transmission, he's been around awhile and is going to be a "dropper" you can trust. If the guy comes up with a crackling, two octaves higher voice, he's either a nerd or so excited he's probably not going to be very accurate. The F-4s checked in with snake and nape . . . perfect. Lead's transmission was good and steady, but Two's was a little unnerved. I quickly briefed the two on the situation, gave them a run-in heading, and while they were flying into position, told Snub Nose to get his team down behind the four-foot drop off where he would be safe from the nape, shrapnel, and debris from the bombs—a perfect granite foxhole. Lead rolled in perfectly on heading. I cleared him to drop two canisters of nape right at the base of the red smoke and then froze in horror . . . it looked like he had dropped the things right on top of Snub Nose. After they hit, right on target, Snub Nose came up screaming, "Stop, stop, you are going to kill us! Stop, stop, you're too close, stop!" I called him, settled him down, and asked if anyone was hurt. I knew their position was located so they would be protected, even from the nape . . . scared yes, but not hurt. He finally settled down, had taken no friendly fire, and when I asked him if he wanted to continue the mission, he jubilantly answered, "You bet, man, I'm writing home about this one. This is really great!" The drop was perfect. Close? How far can you throw a grenade?

I had put the F-4s in orbit, so I called Lead down again to put his snakes just outboard from his nape run. This guy was good. I put Two even out further away from the team, then put the last of their ordnance on the north and south ridgeline. I must say, Snub Nose really had a ringside seat. As the F-4s headed home, I gave them a good BDA, and greatly thanked them for Snub Nose. As a little insurance, I ran the Hueys on the position.

Within minutes, two Marine CH-46s came up on frequency—our choppers to extract the team. Snub Nose popped a white smoke grenade to identify their position and provide wind direction to the choppers. The lead chopper came in, tried to land on the granite rock, but the smoke from our attack and the smoke grenade obliterated the landing site, so he couldn't sit down. I asked him if he could come to the eastern side of the cliff to get away from the smoke and place his nose up on the granite ledge, open his front side door, and load the team. Man, what a beautiful sight! Here's a big ol' CH-46 sitting over a cliff that dropped off straight down, at least 700 feet, its nose wheel sitting on the granite edge, holding so steady . . . it was like a rock . . . great flying, a good driver. The team was loaded quickly and transported back to 3rd Recon.

I later learned that Snub Nose had been through hell up on that little outcrop. The team had been dropped into the site four days earlier and had taken intermittent enemy fire. Someone had run friendly arty on top of them. Taking wounded, they were extracted, the team reconstituted, flown back in, and were about to be overrun when I found them . . . so I guess they were a relieved bunch when we pulled them out.

When I landed back at Quang Tri, I was told that one of our Hueys had crash landed just south of Bia Long, reason unknown. The crew had been rescued, and some friendly troops had been dropped off in the area for security. Now, anytime a Marine Corps bird crashes, people want to know why. Some of the Hueys were getting a little long in the tooth, and if metal fatigue or a part failure occurred, they all might have to be grounded. The powers that be wanted this bird back in the nest. (We could always cannibalize the thing for spare parts.) Seems they were looking for someone with a little authority to take a maintenance crew to the site, inspect the Huey, see if it was salvageable, and, if so, hook it up to a CH-53, lift the carcass out of the jungle, and bring it home.

I guess I was the only guy around who looked like an "authority," so I volunteered to go out with the crew. We all boarded a CH-46 and were promptly dropped into the tall buffalo grass behind the crashed Huey. We were met by a young 1st Lieutenant and a rather brash T/Sergeant (T/Sgt.). The T/Sgt.'s first words were, "Major, what in the hell are you doing out here? Don't you know this is Indian country? The goddamn Huey is a crash. This is bad country. Let's get the hell out of here." I didn't blame him, but we needed that chopper. The pilot had done an excellent job or had been damn lucky to settle her down in the rugged terrain. The Huey had landed, bounced up onto a small hill, and settled next to an enormous tree. The chopper had missed it by whiskers! That tree was dead, at least 100 feet tall, and stood there naked, except for a few bare limbs, some of which made shade for the Huey. The damn tree was obstructing the attempt of the CH-53 to pick up the downed one. The thing had to go.

One of the Huey's rotor blades was laying a little limp. The cockpit, undercarriage, and passenger compartment had survived, and the bird looked in pretty good shape. The maintenance crew gave her a quick look over, didn't find any bullet holes that might have caused the crash, and started rigging her for lift out. I explained the situation and asked the 1st Lieutenant if he had any dynamite to blow the tree out of the way. He came up with some little Private 1st Class wearing granny glasses, who looked to me like he was the last guy in the platoon and had to carry the dynamite. Turned out he was an engineer/dynamiter who won my heart.

We hustled to the base of the tree, which was located on the bank of a small creek several feet below the chopper. Lord, was that thing a monster! My dynamiter felt he could knock it down and away from the chopper. That was good enough for me. While he prepared the charge, I walked back to the T/Sgt. who said, "There ain't no way in hell he's going to knock over that tree." The little PFC soon poked his head out of the brush, received my permission to blow her, gave a resounding, "Fire in the hole!" and whammo! The noise was loud enough to bring every gook in the area down on our heads. The tree shook, shuddered, and grunted, but held fast. The T/Sgt. boasted, "See I told you so. Let's stop this whole thing and get the hell out of here."

I ran back down to the base of the tree. That old rascal must have been centuries old and made of the hardest wood in Vietnam. The blast had only cut the thing about three quarters of the way through! I asked the PFC if he had any more dynamite. He had only used about a third of it. I told him to wire the thing up again. Since we were going to be lifted out as soon as that tree was blown, he wouldn't be needing the rest of his dynamite, so I said, "Don't skimp, put all the remainder on the thing." He did. I stood there and watched him to make sure. The PFC gave his "fire in the hole" and lit the wad. That ol' tree rose about six inches in the air, took a bow, tipped away from the chopper, and came grandly crashing to the ground. Great work! I nudged the T/Sgt. and said, "Any questions, Sarge?" He mentioned someone in the Trinity, turned, and walked away. The little PFC came over to us with a great big ol' shit-eatin' grin on his face. He was one proud pup. The lift chopper saw the tree go and came into a hover over the Huey. The crew hooked her up and they were on their merry way. We called in the circling CH-46s and everyone loaded up and departed the area without incident, but not before I let that little PFC know what a great job he had done.

As I walked into the Ops Office after we landed, I found out our Lt. Parker had been hit by enemy fire up on the DMZ, had been wounded in the arm, had flown back to Dong Ha, and was at the hospital. Later that night, to our relief, he returned to the squadron with his left arm bandaged. He wasn't exactly sure what happened, but with his narration and my detective work, I figured it out when I went up to get the O-1 the next day. The little bird had two bullet holes in its left side. One had entered the cockpit through the left window, shattering it, and proceded on a trajectory above the pilot's head, where it hit the top of the cockpit and fell harmlessly to the floor. A second bullet came crashing through the door, hit Parker's wrist, which was grasping the throttle, and then somehow burst open his wrist watch . . . with bullet, Plexiglas, and debris falling all over the floor of the cockpit. An inch or two either way might have resulted in the loss of the whole package.

I picked up all the remnants, stuck them in my pocket, fired up the O-1, checked it out to see if it was flyable, and returned the wounded little guy home. Later, I took the fragments into Quang Tri City, had them mounted on a squadron plaque, and we presented it to Parker at an O'Club party . . . a nice remembrance.

Chapter 11

Last Days

One March afternoon, I was reconnoitering in the area just north of Khe Sanh when I received a whispered call from a recon team member. He was watching three gooks relaxing in the sun near a large tree on a hill. I put in a call for air and headed for his position. Knowing full well our endeavor might go for naught if I flew over the gooks or fired a marking round, I stayed away from their little sanctuary. The recon kid was great. He talked me right over his position, across the valley from the unsuspecting gooks, by voice command and then described my target: a large tree off my left wing. The tree was the only one sticking out of the jungle within a mile, so it was easy to spot. Two Navy A-7 Corsairs with 500-pound bombs aboard soon became available for our target. I had never controlled A-7s before, but I assumed they would make low, flat run-ins like other aircraft operating south of the DMZ. I gave the Navy guys the required reading. They quickly picked up the target on my voice description alone and set up for their runs. I told the flight to hit just the base of the tree. The gooks were still unaware of the impending doom—one even stood up to stretch. Lead rolled into his run at a very steep dive angle. I was surprised but soon ascertained the flight's experience had been mostly up north, and he was making a normal NVN-type attack. I cleared him to drop. Perfect, right on target! Before Two rolled in, the recon guy came up on the air laughing. "Did you see that? Did you see that? That bomb hit the guy standing up right on the head! He didn't even look up to see it coming!" Our stealthy approach and the Navy steep dive angle caught the gooks unaware of the raining steel.

Mike and Ken had briefed me on one little trick the gooks used to get you off their back. Caught out in the open, if they heard your aircraft, the little guys would drop to the ground in a grotesque position, feinting death, hoping, I guess, that you would consider them dead bodies and fly away. One day with Luckenbaugh as my backseater, we were flying along a small dirt trail north of Khe Sanh. As the trail came to the top of a rise, I spotted five bodies lying on the broadened trail. They were uniformed NVA regulars. I pointed them out to the backseat, who immediately started a request for arty. I called for air, knowing deep down in my heart these guys would be gone before either was approved. Neither was available, and we were left with our own resources. I wasn't about to let the little buggers get away with their possum trick, so I set up in a tight orbit around their position. Backseat unlimbered his rifle, and we planned our attack. We would wait these little guys out, just to let them know we were on to them. Then, when detecting the slightest movement, I'd roll in with a Willie Pete and, as we passed over the gooks, backseat would take them under fire with his M-16 "howitzer." What an imposing force! Finally, tiring of the game, the gooks slowly (and, I felt, sheepishly) started to stand up. I rolled in, fired a rocket, and missed, leaving it all up to the backseat. Not sure he hit anything either . . . but we had "made our day!"

There was one dead giveaway, however, that the gooks never seemed to learn. If they were walking down a trail, for instance, and heard the noise of your engine, they would simply duck down and freeze in position, hoping you might not notice them. However, as you passed over, usually one or two just couldn't resist the normal act of glancing up at you. Believe it or not, that quick little movement drew our attention.

As I made my usual daily sortie up along the DMZ (March 4), flying north of FSB Sierra, I received a call from a recon team saying there were gooks in a couple of foxholes on top and around the sides of the knob. I flew over them and spotted two black-pajama types in one hole, one in another, and several around the perimeter. Arty and air were unavailable, but I lucked out and raised a fellow OV-10 on the radio. He wasn't busy. Each time I flew over the gooks, they just hunkered down in their holes . . . guess they figured I couldn't see them. The OV-10 flew over the position and spotted the holes, so I didn't even have to mark the target. As I cleared him in on his first run, you should have seen them scatter. The OV-10 driver, Major R. G. Whaley, had to make a steep dive to get his rounds down into the holes, and as the gooks heard his cannons bark, two of them flew out of the holes and literally fell down the mountainside. One didn't make it. We put several gun and rocket runs on the hillside and pretty well raked the target. As usual, though, I saw people run for their lives toward the triple jungle canopy and disappear. This was the first time I had really sat back and watched an OV-10 make runs. Compared to the speedy jets, this bird was s-l-o-w. I called Whaley, "Man you guys are really slow in your runs. You look like a damn kite hanging up there!" (The OV-10'ers didn't think much of that statement.)

I thanked the recon team and the OV-10 driver, left the area, and headed south toward FSB Aragone. Approaching the hilltop, I noticed a Huey sitting in the landing zone and idling, from the looks of the rotor blade. On closer inspection, the cockpit was deserted. I couldn't see the chopper crew anywhere. Circling the area, I spotted several black pajama types just down from the crest of the hill. When I called for air, I was told to leave the chopper alone and just put my air all around the sides of the LZ. I ran three flights that really tore up the area. I observed our attack aircraft taking ground fire and was rather surprised. One pilot reported he had seen a machine gun firing at my airplane on one of my marking runs. He rolled in and put the thing out of commission. No aircraft were hit by ground fire during the whole incident.

I was later told by the G-2 folks (true or not) that a battalion commander, one of those ego types who wanted to make a name for himself, had picked what he figured was a patsy landing zone, planned TV coverage to record the deed, and got his ass shot for his trouble. Seems the commander, his S-3, and TV had been in the Lead chopper, in advance of a large heli-lift into the area (without an LZ prep). Just as the bird settled on the ground, they took ground fire, killing all personnel aboard. I guess the gooks didn't know how to shut the whirlybird down, so it was just sitting there running at idle. A couple of escort Hueys had driven off the gooks long enough for another chopper to land, pick up the bodies, and get out. After I left, other O-1s ran numerous air and arty missions around the chopper, but try as they might, every time the attempt was made to put a Huey crew into the LZ to check the bird, they got shot out of the zone. The lonesome old Huey finally ran out of gas and stopped of its own accord. I believe it was later retrieved or dismantled by our own bombs.

March 8 . . . a good day. Lt. Webb was in the back with a handheld camera, and I, being an old photo pilot, expected to take some nice pictures. We were sent out to Lang Vie to photograph what the G-2 folks thought were crates of rockets located next to the burned-out tank hulls that had been sitting there since Lang Vei had been overrun during the Khe Sanh siege. We took some pictures of the new locations, but when I really got down on the deck, all we could see were some flimsy, crating-material boxes, similar to water-logged plywood. We figured if the crates were left out in the open like that, it couldn't be very valuable stuff. The photos didn't come out (had to be Webb's fault—old photo pilots always take good pictures), but during our debrief, we were told that a photo bird from Danang was going to take some photos the next morning. Now, Webb and I both felt the new crates were a set up and possibly put there to suck some unsuspecting aircraft into a triple AAA trap. We highly recommended that the unarmed photo bird be escorted by some aircraft that could retaliate if the photo pilot took fire. (Had to protect my ol' photo buddies back at VMCJ-1.) Sure enough, the next day, the photo bird took fire from the Co Roc area. He didn't get hit, but the scooters that were with him . . . made their day.

After Lang Vei, we flew down around FSB Saigon. By now

our Marines were pretty much in control of the A Shau, so I figured enemy supplies must be coming across the border somewhere between Co Roc and the A Shau. Sure enough, about midway between the two areas, I spotted what had previously been just a small trail. It was now a well-worn, heavily used dirt road. While flying east plotting the road on my map, I noticed another prominent trail leading south. I decided to check it out, and about two miles below the road, while flying about 100 feet above the trail, I spotted slight movement off my left wing. The pattern of objects on the trail seemed to be similar to troops marching as in a parade. I really didn't know what I had, but I whipped my bird into a quick 180 and told Webb to get his camera ready and start taking pictures of the trail as there was something unusual down on the ground. As we flew back up the trail, I spotted what looked to be a company of NVA regulars right out in the open, kneeling down on the trail.

Reaching the end of the column, I climbed for a little altitude and asked Webb if he had seen anything. He hadn't. I made another quick 180, headed back down the trail, and didn't see one darn thing! The gooks had disappeared. I couldn't believe my eyes. Were they really there? Was there anything on that trail? I was questioning my sanity when I suddenly spotted movement in the trees. This was not a heavily forested area, but it had enough trees and bushes to hide the enemy. They were down there. We had struck gookland! I guess what made the situation so unbelievable was that I had never seen this many enemy troops out in the open before. I immediately called for air . . . no, arty . . . no, nothing available! We flew back and forth taking numerous photos of the area (none came out). Damn, here was my opportunity to create a little havoc in the war, and I couldn't get round one on these guys. This trail led right down into the A Shau, and apparently these troops were on the march out of the A Shau when we happened along.

I kept circling the area just to make them sweat in the hot

sun. They were probably digging a lot of foxholes. After some thirty minutes, air/arty were still not available, so we went back to the east-west road. We called every fifteen minutes to check on our requests; neither came through before we had to head for home. The gooks wouldn't need their foxholes after all. One spot on the first road really caught my attention. It crossed a small stream about six miles in from the border. There was a graded parking area on both sides of the stream.

Arriving back at Quang Tri, I asked the S-2 folks if there had been any sightings on this new road. They didn't have anything, but Webb went over to Recon and found out we had a recon team on a hill overlooking the road and stream. The team reported truck traffic using the road for the last few nights, frequently stopping at the stream to fill their radiators. They had seen the trucks roll east at about midnight every night, then return west in the early morning hours. The gooks had been shut out of the A Shau by our Marines, and were now routing supplies and troops along this new road.

Previously I heard through the grapevine that our A-6s had a new-fangled capability: put a black box in the hands of a FAC on the ground, and he could then point the thing at a ground target and direct the A-6s smack dab on top of that target. I told our S-2 shop it might be a good idea to check out the A-6 capability up through the chain. If it existed, a FAC with his black box should be flown into the recon team's position and let the A-6s have a field day. This was done, but rumor had it . . . when the birds arrived overhead, the box didn't work. The FAC radioed for a TPQ mission, but by the time it was worked out, the A-6s were short on gas and had to head home. They tried again when the trucks headed back west, but fog set in at Chu Lai, and the birds couldn't get airborne. The teamwork didn't work this time. Additionally, I flew over the area several times later and never spotted any signs of an attempt to stop the enemy from using the road . . . a disappointment!

Just a word about the A-6. Although I believe most of the A-6 drivers were former A-4 pilots, they didn't seem to be as accurate as the good ol' scooters. Nevertheless, they carried so many bombs, you could spend half a day trying to get it right.

Went back to "my road" in an OV-10 (March 9) and spotted a couple of workers on the road, but before I could swing into line to take 'em out, they disappeared into their roadside tunnels. I spent all my ammo shooting into the several tunnel entrances along the road, but I must say, old photo pilots don't make very good attack pilots. This was my last OV-10 hop in Vietnam.

One day, our S-1 folks briefed us on a Marine company-size troop maneuver that was just kicking off. The company was moving out of Vandy, north around the Rockpile, up the valley west to Khe Sanh, then north to the DMZ. I followed their movement for about a week. These poor guys had hiked and hacked their way through some of the roughest terrain in the world. I don't believe they took one step out in the open. Once they reached the Khe Sanh plateau, instead of traveling over brushy, open areas, they stayed in the confines of the jungle and endured its many problems. I just couldn't imagine how they could possibly navigate in that mass of foliage. We hadn't spotted them for about three days when, while flying up along the DMZ, I got a call from their leader. He had spotted me and directed us over his position. Upon arrival, I saw some pooped out troops.

To digress again . . . the Marine Corps had a classified coding system we used when talking to the grunts, specifically when the conversations involved map coordinates. These were set up on a day-to-day basis. Basically, we could transmit encoded coordinates over the radio to ground troops so the enemy, who monitored our frequencies, couldn't recognize where our troops were located and bring weapons to bear on them. We seldom used the codes (most of our coordinate calls were on enemy positions) but were issued same for this operation.

As suspected, the company was lost. We found them strung out in a deep canyon on the headwaters of the Ben Hai, spreading north up a ridge to the top of a mountain about 1,000 feet high and then along a ridgeline just into the northern part of the DMZ. They had hit the Ben Hai, apparently without water, and were passing water canteens back and forth to the river. These guys had to be in dire straits. Their leader queried his position and gave us his coordinates. My backseat broke out the code book, only to match up a location some three miles from the leader's location. We radioed back his correct position, which he just wouldn't believe. After about ten minutes of checking and rechecking to see if we were using the same codes, he asked us to come out in the open and transmit his coordinates without using the codes. We dropped down to the canyon floor in an attempt to screen our transmission from preying ears and gave him the correct location. He was aghast, and I don't blame him. He had reconnoitered 40 to 50 miles of some of the toughest terrain in the world and, in my opinion, had the right to be lost.

Boy, my heart went out to these guys. Their leader's voice transmissions were becoming rather forlorn when he finally asked if we could get him some helicopters to transport them out. I found an LZ about 100 yards downstream that looked large enough to hold one chopper. I figured the troops were in such a remote location that there was slight chance of any enemy being around. While talking over the situation, right out of the blue, one of our Hueys came wop-wop-wopping up the canyon. He came up on the air and said he had been told to go out into the area, find the company, and determine their situation. I recognized the Huey driver's voice as one of our squared-away pilots. An expert had arrived. He figured a CH-53 could slip into our picked LZ, and although we were a little late in the afternoon, he could probably get them all out. Choppers had already been put on standby to assist him if needed . . . good, prior Marine Corps planning. Knowing the situation was now in capable hands, I left the area. Later, after returning to Quang Tri, I found out the last chopper load came out of the canyon just as the veil of darkness settled over them.

One day, I was flying way out on the Laotian border without a backseater. Now, I had flown over a lot of mountainous terrain in the Bird Dog and never had any trouble with downdrafts. I thought I saw some movement on the other side of a small ridgeline, and as I flew over the thing at about fifty feet, I became a passenger! Something reached out and grabbed that old O-1 and started us down into the trees. I immediately added full power and wiped out the cockpit with the stick, but neither had any effect. Man, I was headed for the trees and I couldn't do a damn thing about it. My next thought was . . . Christ . . . if I survive this crash, I'm going to be one lonely SOB. Our people would never find me out here in Laos. God must have heard me swearing and wanted to save me for a fate far worse than this impending disaster, because suddenly I was in control again. I recovered about ten feet above the ground between some tall trees that I swear should have taken off both wings! What a "pucker"!

March 15 . . . A bad day weather-wise. Quang Tri had about a 500-foot overcast, but figuring I might find a hole to get out into the area, I saddled up ol' Bird Dog and headed up the road to Dong Ha. The low clouds around Dong Ha were around 200 feet, but as I proceeded up Highway 9 towards Vandy, I ran into a solid wall of clouds before reaching Cam Lo. On the return route, I noticed a Navy ship headed out from the docks on the Cua Viet river, just east of Dong Ha. What really caught my attention were several kids throwing rocks at a bunch of sailors on the fantail of the ship. The ship's crew were returning fire by heaving apples, oranges, heads of lettuce (in other words, garbage) back at the kids. This didn't seem to be a peaceful endeavor—the kids retrieved the garbage and heaved it back at the sailors. I thought I could break up the melee by swinging around behind the kids, making a low pass just over their heads, and scaring them off the fight. As I started my roll in, the ship was getting out of rock range, and as the kids gave up and turned away from the ship . . . all they could see was one olive drab, green Bird Dog coming right down their throats. Man, did they scatter! I guess this rock-throwing contest got to be a normal thing, and apparently the Navy was getting tired of the fight. As I flew over the ship, the white hats gave me a "winners" wave!

The officers' comfort station on Base was a three-holer that rested near the edge of and with its back toward the Quang Tri River. It sat there well within rifle range of any enemy wanting to hose it down from across the river. In fact, Fritz's jeep took a round in the back one night when it was parked nearby. One of our piss tubes was located just out back of the Majors' hut, a much safer area. One night, I walked out to gain a little relief and found out some enterprising individual had jury-rigged a florescent light on a wooden frame just above the tube. Remarkably well constructed, it even had a light switch and a wire leading into the ground. Several voiced their opinion that the lighted tube was even more dangerous than the three-holer. The discussions revolved around the possibilities of being electrocuted by the confluence of the underground wire and the dampness caused by the big pee-ers and bad aimers amongst the users, or the possibility of being an attractive target when you flipped the switch on at night. Of course, it was strictly a humorous setup, and it was amazing how the secret was kept. (I'll admit I tried the light switch my first time at bat. It didn't come on.)

One night a young boot 2nd/Lt. came by. Someone had reported a malfunction in the Piss Tube Lighting System to Base Maintenance, and they sent the boot out to see what was wrong. A bunch of us were sitting out back when the guy came up and had a great deal of conversation about what the trouble might be. The boot assured us he would have a crew out at 0700 the next morning to dig up the wire and trace the agreed-upon short in the system. This was just too good to be true. The next morning, there was an audience waiting for the "Piss Tube Fixers." The lieutenant showed up right on time with three enlisted troopers, shovels in hand. He walked over to the light switch to demonstrate the malfunction, and although he didn't know the route the wire traveled underground, he instructed them to dig out the wire to have a look see. With a look of "I can't believe this shit," one of the enlisted men reached down and jerked about three feet of wire out of the ground and said, "Lieutenant, here's your problem. It's not hooked up to a damn thing." To appreciate the incident, you had to see the look on ol' Lieutenant's face when he saw the end of that wire. He was so flabbergasted, he couldn't even speak. We all had a big laugh at his expense, but he didn't have to buy one drink at the club that night.

One day we were flying out west of Lang Vei, just on the Laotian border north of the Co Roc cliffs, when I spotted what appeared to be a dust storm being created by a large helicopter either landing or taking off. It was about ten or twelve miles on the other side of the border and was in a highly NVN built-up area. I was sorely tempted to fly over and try to score one for the O-1s by hitting the thing with rockets. What a story I would have to tell if it worked; however, the longer I thought about it, the smaller my balls got. I would be walking a thin line between good sense and being crazy. We didn't have arty within reach of the blasted thing, so I called for air; none available. (Man, this could get frustrating at times.) All I could do was call in the target and hope air became available. It didn't. Back at Quang Tri, I briefed our other pilots and suggested to our S-2 boys that a TPQ or Arc Light be put on the area. That's the last I ever heard about it. Might have been an Air Force or Army chopper . . . NAH!

Just back from a morning hop, I was sitting in the Ready Room when Dub asked me if I wanted to go with him up to the Rockpile. Now, the Rockpile was a high, small, mountain peak with sheer cliffs dropping some 700 feet to the valley floor. It had a little ol' landing platform right on top, and I assumed it was one hell of a thrill to land on and take off from the damn thing. I really believe the invitation for the hop was to scare the hell out of me. We had Marines living on top of this edifice whose only source of resupply was by chopper.

Dub had a recon insert down on a highway below Co Roc, and after the team was safely on the ground, we headed for the monster. Ol' Dub was a hell of a pilot, so I had no qualms about landing on the platform . . . sure did look little, though . . . and high. The north end of the landing pad dropped straight off onto the valley floor. The landing was uneventful, except for the fact our rotor wash nearly blew the troops and their shelters off the top of the mountain. We shut down and got out to one of the most magnificent views I've ever seen from the ground. Fantastic! What an observation post. I didn't really know whether or not to believe some of the troops I talked to, but they told stories of the other inhabitants of the peak: rock apes (large monkeys that frequently got into rock fights with our troops.) Scared hell out of them when a rock ape tried to crawl into a fighting hole with them or a sleeping bag. They were said to be rather ferocious. The gooks had tried to take the Rockpile numerous times, but always failed.

We spent an hour exploring the place and talking to the troops, and believe me, I didn't get over an arm's length away from Dub. There was no way I was going to let him fall off that mountain, leaving me to fly that damn Huey back home. The takeoff was quite a thrill—just dropped off into space—but I sure didn't let on to Dub that I gulped once or twice looking over the edge of that cliff. (I always wondered what it would be like to shoot a touch-and-go landing on the small landing pad in an O-1).

I was assigned my first convoy guard hop on March 17. The grunts liked to have O-1 or Huey air cover for their truck convoys leaving Dong Ha on the trip up to Vandy. Usually an uneventful hop, it did let you flat-hat and show off around the trucks.

The next day, another convoy hop! This was not normal. We let the new guys fly these hops because they were safe, and it gave them a good chance to feel out the Bird Dog. Then it dawned on me—someone back in Ops was scheduling me for the easy hops because I only had ten days left in country. Now, a few pilots in NAM got "go home-itis" and protected their ass by not flying the last month of their tour or asking to be scheduled for easy hops. God, I wasn't going to be one of them. I had heard accounts of pilots getting too cautious just before they went home and getting killed by "safe-in' up." I had told the folks in Ops to keep me on a normal schedule, and here they were protecting me. Maybe this is my real braggin' time, but a couple of months earlier we lost a Marine F-4 while on a close air support mission up on the DMZ. He ran into a mountain. A good friend and old VMCJ-1 squadron mate was sent up to investigate. Before he left, he told me the loss was an accident and that our Bird Dog guy who was controlling the flight was not to blame. Then he said, "Lloyd, you've got quite a reputation with your fellow pilots and AOs. They say you're a little reckless and take a lot of chances, but the AOs love to fly with you . . . and they say you're the one guy who will go anywhere, anytime." Lord what a feeling!

Well, now, a guy with a rep like that couldn't start taking those "pussy" hops, so I told the Ops folks to delete me from any further convoy missions. However, things did get interesting during that hop. As I was flying over the north ridgeline just west of Cam Lo, I spotted a foxhole manned by a gook overlooking the convoy with binoculars. As I flew over him, he crouched down in the hole, trying to make himself invisible. I called for air—no dice. I didn't have a backseater, but I took out a smoke grenade, just to let the gook know he wasn't alone. He skedaddled for a treeline before I could drop the grenade. (If he only knew how accurate I was with the grenade.)

When the trucks were safely sequestered at Vandy, I headed for my favorite road and spotted some gook workers repairing a bomb crater. They dove into their little roadside tunnels when they saw me. I called for air, and in a few minutes, one of our OV-10s came along.

He wasn't too accurate, but we really worked over a bunch of tunnel entrances with his rockets and mini guns. How many gooks did we kill? Who knows? You can't see down into tunnels.

On the way home, I flew over my little friend in the foxhole. It was empty. I circled the area for a few minutes looking for enemy troops around the foxhole but never came up with anything. On return, I briefed the S-2 folks and suggested someone better put some steel on the area. A couple of days' later dispatches indicated we had put troops into the area, and about 100 yards from the foxhole, they made contact, killed several VC, and destroyed a small cache.

Today (March 19), I had probably the best seat in the house to watch our Marines making war. Flying the Bird Dog was the greatest experience of my life. You could see the war going on right in front of you, the troops advancing up a hill, the gooks entrenched, a call for air or arty, see the results, watch our troops advance and achieve their objective . . . better than any movie . . . and proud of your contribution. I received a call to assist our grunts on a ridgeline just northwest of the Rockpile. The war had already started. When I arrived, our troops were pinned down. I guess the gooks thought they were safe in their trenches, because they were easily spotted from the air and didn't seem to want to hunker down. Our grunts were lobbing some mortars onto the hill but didn't seem to be having much effect.

I knew the FAC on the ground, an OV-10 pilot from our squadron. I gave him a good brief on enemy positions and was arranging for air when a CH-46 came lumbering into the area. I had already taken fire from the enemy hilltop and tried to tell the driver that he could expect ground fire. Luck wasn't with him this day . . . he got shot down. He was, however, lucky enough to auto rotate into a landing site—right on top of the enemy hill. As they hit the ground, the whole gaggle dove out of the bird on the opposite side of the hill from the enemy except one guy who jumped out the wrong side, right into the enemy's lap. I heard later that he was a M/Sgt. who liked to go along on hops as a door gunner. Guess he was a little startled when he looked around and saw all those unfriendly faces. Intense fear drove him back up to the chopper and down the proper side to be with his buddies.

Air came online about then . . . A-4s. We dropped nape and snake all around the hill. Another chopper flew in to rescue the downed folks, took some ground fire, but flew out without a scratch. I then ran two more flights on the place. When our troops attacked, they met only light opposition.

March 20 might have been the luckiest day of my life. But first I have to explain the two arming switches on the O-1. These two switches were awkwardly located on the ceiling of the cockpit, just to the left of the pilot's head. The trigger was on the stick where most respectable triggers are. Before firing, you had to uncover the arming switches by clicking the red safety switches back and then you could toggle the proper arming switch for the rocket you wanted to fire.

On my morning hop, I was asked to help some of our Marines receiving enemy mortar fire. Once in contact with the "receivees," I was given the location of the weapons firing at them. Over the area, I spotted some gooks on top of a heavily forested hill. Air was immediately available—A-4s. I had flown rather low over the site and hadn't had time to climb back up to a reasonable run-in altitude. As I rolled in on my marking run, I flicked back the red cover, levered the arming switch for the left rocket pod, lined up, hit the trigger, and nothing happened! I frantically repeated the arming procedure . . . still nothing! By then, I had to pull out at treetop level and really got hosed down by some automatic weapons. I called the A-4s, telling them to orbit until I could get into position for another marking run.

When in my second run, I flipped the switches for the right pod . . . nothing. Recycled . . . nothing. I armed all the switches again, sighted in, and not a damn rocket fired. We again received ground fire, and as I pulled off target at tree top level, the ground fire sounded intense! Coming around for another run to mark the target with a smoke grenade, my backseat mentioned the fact we had taken ground fire all the way down the run. I hadn't paid it much attention—too busy. (I was still invincible.) As I came around for my next run, bless his heart, Lead said, "Seaworthy 14, you don't need to make another run—we can easily identify the target by the flash of weapons that were firing at you!" The A-4 attack was a thing of beauty . . . and a great feeling of revenge from inside my cockpit. I'll never know why we weren't blasted out of the sky.

I was only five days away from my going-away party and six days from leaving Vietnam. With no specific assignment for my afternoon hop, I decided to just fly around the area looking at the beautiful scenery. One of my favorite spots was a group of granite mountains located way up in the northwest corner of I Corps, just below the DMZ. Lt. Webb was my backseater, and as we were tooling along over the granite, I heard automatic weapon fire from our six o'clock position. Webb saw the tracers flying up at us and said they were low and to our left. He spotted the location of the weapon, and I immediately called for air. I marked the target, and we really blasted the neighborhood.

After the air strike, I thought to myself . . . Draayer, ol' friend, here you are with five or six days left in NAM, you've just had two rip-roaring hops, what's next? I soon found out. On the way home, just northwest of the Rockpile, I observed a gaggle of Hueys approaching an LZ. As one Huey touched down, they started taking ground fire. The escort Hueys rolled in to stop the fire, and the next thing I saw was the Huey pulling out of the LZ . . . with a Marine radioman hanging onto the skid! At about 400 feet above the ground, the Marine's legs seemed to spasm and he let go, plunging to his death—a heart-wrenching tragedy.

After two "shot-ats," being "short," and having witnessed the terrible loss of a fellow Marine, I thought, ol' man, maybe it's time to hang it up. It was. When I pulled into the chocks and shut the little bird down, I suddenly realized . . . this was my last hop in the O-1. It brought tears to my eyes. I really loved the little guy and the FAC mission. (As a rodeo clown, my motto was "You always get the best seat in the house and never have to pay to get in." This was certainly true in the O-1!)

Later in the evening, Dub told me the full skid story. For a couple of months now, we had used Army slicks (Hueys without guns) to transport our recon teams into their sites. When our Marine slicks were used to insert recon teams, if the team took enemy fire, the Huey escorts immediately dove on the enemy to neutralize the situation. Our slicks either remained in the landing site until all team members were back aboard or, if airborne, immediately dove back in to load the team and get out of the area. We had already had a couple of instances where the Army boys, receiving fire, got the hell out of there and abandoned the recon team. The only saving grace was when our escorting Marine Huey gunships hosed down the area until it was safe for the Army to go back in. One recon procedure was mandatory: the man with the radio on his back was the first man to exit the chopper and the last to board when they were picked up.

That afternoon, Dub's flight was escorting two Army slicks into the site. As the first slick went in, the radioman and another trooper jumped out. The pilot heard gunfire, and before the two recon guys could climb back aboard, he started to lift off. One team member managed to jump up into the chopper, but the radio man came away . . . hanging onto the skid. Dub screamed for the slick driver to get the chopper on the ground, but the numbnuts continued to climb. Somewhere around 400 to 500 feet above the ground, and despite attempts by his other team members in the chopper to haul him in, the radioman lost his grip and dropped to the ground. We found out later the recon team leader had pulled his .45 pistol, held it to the pilot's head, and threatened to shoot him if he didn't get that chopper on the ground. The pilot seemed to be in a state of shock and didn't comply. The Marine's body was later recovered.

A little later that night, when Dub found out I had flown my last hop, he said, "Draayer, you don't have a hair on your ass if you don't go out with me tomorrow!" I immediately accepted the challenge. This put Dub into a frenzy. "No, no, NO! I was just kidding. I can't take you with me. You'd get us both shot down. Man, with that lead-drawing ass of yours . . . I just can't let you go with only a couple of days left in country, there's no way." I vowed to accept that challenge and called Ops to schedule me with Dub the next morning. They didn't. I had a plan, however. I hid from Dub all morning until he went into the line shack to sign off his Huey. He had sent his 1st/Lt.-type copilot out to preflight the aircraft. I chased him away, telling him about the challenge. I took the copilot's seat, put on my helmet, pulled the dark sun visor down, and waited for Dub. While he was preflighting the Huey, he kinda kept glancing at me in the cockpit. Finally, when he got in and recognized who was sitting in the copilot's seat, he about went into shock. We had a little discussion about me being there . . . just a little one, because I outranked him. He fired up the bird, and we were on our way.

The mission for the day was two recon inserts. The first one was uneventful. The second one was way up on a mountain some folks called Tiger Tooth. This thing was so far away from anything that I knew there couldn't be any enemy troops in the area. Wrong! We were escorting two infamous Army slicks and, sure enough, as two team members jumped out of the chopper, the ground fire started. The slick panicked and pulled out of the zone. We spotted smoke from the ground fire, and Dub immediately rolled in and took the area under fire. It was just pitiful seeing two of our Marines laying there, huggin' the bottom of a small bomb crater, with bullets flying over their heads. The firing stopped, and Dub contacted the two on the ground. Luckily they hadn't been hit, so Dub chewed a little Army ass and sent the slick back in. The troops came out.

Dub told me this type of helo-trap happened quite often. The gooks knew where many of our landing zones were and would simply station a couple of shooters next to the zone, allow our choppers to land, and then take them under fire. It was time to go home (after being airborne for just over five hours), so Dub decided to expend our ammo around the landing zone. He even let me take over and fire on a couple of runs. After he chocked the Huey, a plane captain came up to the cockpit and told us we had one single bullet hole in the tail of the aircraft. Three "shot at" hops in a row . . . this was my last hop. If I was going to get killed in Vietnam, it sure wasn't going to be in a damn chopper!

My going-away party was scheduled for the evening of March 25 at the club. I was leaving the next day, so I spent the whole day checking out of the squadron and saying goodbye to the troops. It was really tough saying goodbye to the O-1 maintenance folks. They really worked their butts off keeping the Bird Dogs flying. I went out to the O-1 flight line and, with tears in my eyes, shook each hand.

It was especially difficult to say goodbye to Sgt. C.B. Lockwood, as fine a Staff/NCO as I ever met in the Marine Corps. A shy, hard-working Marine, he somehow kept coming up with parts for the O-1. The Air Force had transitioned over to the O-2 Skymaster; the Marine Corps was flying the last of the O-1's used by Americans in Vietnam; we had the last of the O-1's in I Corps, and they were being phased out in favor of the OV-10. The supply line was practically non-existent. I'll never know how he did it, but somehow Lockwood would either dredge up some orders, or, as I suspect, he just jumped on a C-130 leaving Quang Tri, traveled all over Vietnam, and arrived back with a load of spare parts for the Bird Dog.

One day after he took R&R in Hawaii, he and a couple of other O-1 mechs came into my office. I was always kidding around with them and asked what jail one of our mechs was in now and how much would it take to bail him out. Lockwood had one of those shit-eatin' grins on his face and had his arm behind his back. He brought out a small trophy with a racing type airplane on top. He apologized for not being able to find one topped off with an O-1. The mechs had all chipped in, Lockwood had taken the time to buy it in Hawaii and have it engraved with a most heartwarming inscription: "TO MAJOR DRAAYER FROM THE MEN OF O1-C WITH LUCK." I accepted the trophy, said thanks, then told them to get the hell out of my office because I was fixin' to bawl my eyes out . . . and I did. I really left a chunk of my heart with those guys.

Earlier, Lt. Webb, a recon officer and my favorite backseater, came into the Majors' hut and presented me with an enemy SKS rifle. A recon team had brought it back from an insert-ambush. He said it was my war trophy from recon. All the paperwork was filled out, and he guaranteed I could get it out of Vietnam and home. Believe me, I overly protected that rifle. Not many aviators had access to an enemy piece of equipment, so on the way home, just about every time I had to check it through a Marine checkpoint and customs, I received many looks of disgust from the ground marines who figured I was some RMF who hadn't earned the right. The usual question was, "Where did you serve in NAM?" Their attitude sure changed when I stated I flew Bird Dogs out of Quang Tri. I slept with that weapon all the way home.

My going-away party was a zinger. It was held in the club where the standard Marine Corps party fare was served (potato salad, baked beans, cold slaw, rolls, and a few meatballs). I made sure the club manager had plenty of champagne on hand, gave him 100 bucks to buy the house, but told him not to pop the cork until right after my going away speech. Fritz, bless his heart, sat up at the head table and placed three bottles of champagne in front of me, stating, "Draayer, I've never seen you falling down drunk. Tonight's the night." Well, I gave a hell of a speech and ended up by saying, "Goodbye, you bastards!" I then shook up the bottles in front of me, hit Fritz with the first one, and proceeded to drown the joint. It wasn't long before every one was popping corks and pouring the contents on the nearest head.

We soon got a carrier landing going. Explanation: Line up two or three tables (the carrier deck), get two or three guys to stand on each side of the tables to act as arresting cables, pour drinks on the tables until they are sloppy wet, and then get some guy to take a run at them, do a belly flop, and during the slide, the arresting cables lock arms and try to hook his raised legs as he goes screaming by. We usually missed. If the "cables" miss, he goes flying off the end of the table! Several tried, all missed. Soon someone came up with the bright idea of placing champagne bottles on the end of the tables . . . à la a bowling alley. Little Mo was the first bowling ball. He came away with a cut on the top of his noggin but knocked most of the bottles down. Undaunted, he put on a G.I. helmet and tried again but left one bottle standing as he went crashing to the floor.

About thirty minutes into the party, the barkeep came up and told me he was out of money. I gave him another 100. Not long after, he came to me again; he was out of champagne. I told him to just start passing out wine bottles. I ended up at the end of the bar where I developed some special teamwork with the barkeep. He kept filling pitchers with beer or, if we were in a hurry, water, which I distributed on most of the unlucky passersby! Sometime during the melee, someone came up and showed me a picture he had taken—me sitting with a champagne bottle in one hand raised in salute and three slices of baloney resting lackadaisically on my bald head with an olive stuck on top.

Finally, someone figured the duty helicopter watch pilots were over in their racks missing all the fun. I grabbed a bottle of wine in each hand and headed for the Majors' hut. Major Barkovich was sound asleep when we arrived, but not for long. We drowned him with six bottles of wine. About twenty of us stayed in our Quonset hut where it was a little quieter. The party at the club got worse.

It seems a Huey driver started lobbing rolls at the CH-46 crowd. They retaliated. Soon they grew serious and started heaving the things like bullets. Run out of rolls—just use handfuls of potato salad or baked beans . . . the meatballs were all gone. I really don't know how that party ended, but it seemed to end up at our hut. I went to bed, praying to be left alone—there were a lot of animals roaming around Quang Tri that night. Most of them paid a visit to our hut and emptied sandbags all over our kitchen floor. There was a good four inches of sand on the floor when we finally rolled out of bed the next morning.

Fritz soon received a call from the group commander. The conversation:

Boss: Have you seen that club this morning?

Fritz: No, sir.

Boss: Well, it's a hell of a mess! There's beans and potato salad thrown all over the place, broken bottles on the floor, and some of the overhead fans are torn out!

Fritz: Oh, yes? (Fritz wasn't entirely awake yet.)

Boss: What I want to know is what are you going to do about it?

Fritz: Well, we must have made the mess. We'll clean it up.

It didn't take us long to wake up the other majors and a couple of captains, and head for the club. Lord, was it a mess and did it smell! There were beans on the ceiling. Guys had hung from the fans and dared their buddies to shoot them down with handfuls of beans, cole slaw, or potato salad. Most of the window screens had debris stuck to them, and you just wouldn't believe the amount of glass lying on the floor. We rounded up some brooms and mops and proceeded to clean the place. Before we finished, word spread throughout the camp about the party, and a whole lot of enlisted men walked by to catch a glimpse of the brass conducting a clean-up detail. Most of them left with smiles on their faces and no pity in their hearts.

That afternoon . . . I left Vietnam!

"For those who fight for it,
Life has a special flavor,
The protected never know."

(Pinned to a bulletin board at the
Marine Command Post at Khe Sanh)

GLOSSARY

A-1 Douglas A-1 Skyraider (Spad) Propeller-driven multipurpose attack bomber. Of Korean War vintage, capable of carrying huge bomb loads, armed with four 20-mm guns. It carried enough fuel to remain on station for hours, requirements that greatly enhanced the Jolly Green Giants' rescue attempts for downed pilots.

A-4 (Scooter) Douglas A-4 Skyhawk, the best little bomb dropping platform in I Corps. Flown by some of the best "droppers."

AAA Anti-Aircraft Artillery. Rapid firing cannon or machine guns, often with interlocking fields of fire (also known as Triple AAA).

Abort To suddenly cancel a takeoff, flight, or mission.

ADF Automatic direction finder, a receiver that detects signal strength and provides bearing information from a transmitting ground station or another aircraft.

Air America The name of the CIA's proprietary airline operating throughout Indochina.

Air Medal Awarded for single acts of merit or heroism while participating in aerial flight.

AK-47 Soviet automatic assault rifle used by the VC and North Vietnamese. An effective weapon.

Ammo Ammunition. . . Ordnance.

Arclight B-52 bombing missions.

Arty Artillery.

ARVN Army of the Republic of South Vietnam, the little friendlies.

Auger Loiter.

Auger-in Military pilot's term for "crash."

B-52 The Boeing B-52 Stratofortress, Air Force bomber. Could carry 84 500-lb bombs plus 24 750-lb bombs.

Backseat The ground officer who sat in the back seat of the O-1. Also called AO, backseater, guy in the back, etc.

Ball (Also meatball.) Optical landing aid that guides pilots while landing aboard carriers.

BDA Bomb Damage Assessment. Report made by the FAC/AO team after an air/arty strike, indicating results: troops killed, buildings or trucks destroyed, secondary explosions, etc.

Bingo Military lingo for "short on fuel, time to go home."

Birddog O-1 reconnaissance plane.

Blackshoes Navy surface or ship personnel. They wear black uniform shoes, whereas aviators wear brown shoes.

Bonnie Dick Nickname for the aircraft carrier USS *Bon Homme Richard.*

Buy the farm A favorite euphemism for dying in a crash, "bought a chunk of Vietnam."

Cat Short for carrier catapult.

CH-46 Boeing CH-46 Sea Knight transport helicopter, could carry 25 troops, 4,000 lb. of cargo, or 15 stretchers. Brought troops and cargo on resupply missions to the front lines, many times under enemy fire.

CAG Carrier Air Group Commander. Boss of all air operations aboard a carrier.

Call Sign The identifying letters or numbers assigned to an aircraft for the purpose of visual identification or radio communications.

CAS Close Air Support—any variety of fire support missions prosecuted in close proximity to friendly positions; weapons delivery flown in direct support of ground troops.

Chopper Helicopter.

CIA Central Intelligence Agency.

CO Commanding Officer.

Connie Nickname for aircraft carrier USS *Constellation.*

DASC Direct Air Support Center, the ground-based control center that was responsible for assigning attack aircraft to FACs. Also called ground or control in the text.

DFC Distinguished Flying Cross, awarded for heroism or extraordinary achievement while participating in aerial flight. The performance of the act of heroism has to be voluntary, above and beyond the call of duty.

DME Distance measuring equipment. Feature of airborne navigation equipment which measures distance to a ground navigation station.

DMZ Demilitarized Zone. Buffer zone between North and South Vietnam established by the United Nations in 1954. The Ben Hai river generally meandered through the middle of the DMZ.

ECM Electronic countermeasures, a variety of active electronic jamming procedures.

Elint Electronic intelligence, a variety of passive snooping on enemy electronics capabilities.

EWO Electronic Warfare Officer. Guy in the right seat of the F-10/EA6A—sometimes called R/O.

ExO/XO Executive officer, second in command.

F-4 McDonnell Douglas F-4 Phantom II. Two-seat, all-weather, long-range attack jet fighter; became one of the principal bombers of the war.

FAC Forward air controller, TACA(A), Birddogger. The pilot flying the O-1 who controls attacking strike aircraft engaged in close air support of friendly troops or against other targets. Usually an ex-jet pilot.

Feet Wet Military term used to report when aircraft leaves land (feet dry) and is now over water.

Freq Refers to radio frequency.

Friendlies The good guys.

G-2 Marine Corps Wing Intelligence Office. Also slang for reconnoitering an area for intelligence . . . "We G-2'd the area."

GCA Ground/carrier controlled approach. A controller, using a radar landing system, talks the pilot to a landing. The pilot flies on instruments and follows instructions.

Guard The universal emergency radio frequency that all aircraft are required to monitor.

Gunships Armed Huey helicopters.

Gooks The bad guys.

HE High-explosive ammunition designed to explode on impact.

Huey Most widely used helicopter of the Vietnam War. Fitted with two forward-firing machine guns or launchers for 2.75-inch rockets. M60 or other machine guns were hung in the open doorways, manned by enlisted personnel.

Jolly Green Giant The nickname for the armored Sikorsky HH-3E helicopter, specially modified for search and rescue (SAR) operations and flown by the Air Rescue Service of the USAF. The guys that pulled our downed pilots out of the bush.

KBA Killed by air.

KIA Killed in action.

Kneeboard A device strapped to a pilot's thigh which will hold maps, charts, etc. A pilot's miniature work table.

Lead Pilot talk for the guy flying the lead aircraft in a formation flight of two aircraft or more.

LZ Landing zone, the designated site for a helicopter insert of troops and/or supplies.

M-16 Rifle used by our ground troops and our FAC/AO's in the Birddog.

MAG Group Marine Aircraft Group, an intermediate level of command composed of a variety of combat and support squadrons.

Marston Matting Metal planks used to construct runways on remote strips.

Mayday Distress call made by a pilot when in an emergency situation.

Meatball Term for circular ball of light in the optical landing system reflected up the glide slope; "the ball."

MIA Missing in action.

Napalm/Nape A large tank attached to an attack aircraft filled with fuel that explodes in a huge volume of fire when contacting the ground. Used against enemy troops close to the friendlies.

NAS Naval Air Station.

NVA North Vietnamese Army.

O-1 Birddog, spotter plane used by FACs in Vietnam and Laos.

O-2 Air Force spotter plane that replaced the USAF O-1s. Military version of the civilian Cessna Skymaster.

OV-10 North American Bronco, designed for FAC use; replacement for Marine Corps O-1s. Armed with mini-guns and rockets.

One-eighty...180 One hundred and eighty-degree turn. Reversal of course.

Photo interps/interps Photographic interpreters who glean intelligence information from aerial photographs.

Puff the Magic Dragon Marine nomenclature for the old R4D/C-47 transport aircraft, modified to carry three mini-guns firing straight out from the left side, each with a rate of fire of 6,000 rds/min. They slowly flew left-hand turns around ground targets.

Recce Reconnaissance flight.

RPG VC/NVA portable, hand-held rocket launcher; small, light and with a big "punch."

S-1 Marine Corps Squadron administrative office.

S-2 Marine Corps Squadron intelligence office.

S-3 Marine Corps Squadron operations office.

SAM Soviet surface to air missile. A dangerous weapon.

SAR Search and rescue. An elaborate procedure under Air Force rules involving the suppression of ground fire by fighters before Jolly Green helicopters attempt a downed pilot pickup.

Secondary explosions Explosions following a first attack on a target; ammo/stores that "cooked off" as a result of lingering fires from the initial attack.

Silver Star America's third highest medal for bravery, gallantry in action against the enemy.

Slicks Unarmed Huey helicopters used for troop lifts.

Snake eye/Snakes Bombs designed so that after release, fins pop open to act as air brakes, which allow the bomb to fall far behind the attacking aircraft. This decreases the risk of damage to the aircraft from the bomb blast. Extremely accurate.

Spad Nickname for A-1 Skyraider.

Spooks Air America/CIA.

T-28 North American Nomad, a two-seat armed counterinsurgency propeller-driven airplane, extensively modified for use in Vietnam and Laos.

TACAN Tactical air and navigation control system, developed by the military to enable a pilot to pick up a radio bearing, providing directional and distance measuring information.

Tico Nickname for aircraft carrier USS *Ticonderoga.*

Two Pilot talk for the number two aircraft in a flight. Wingmen were: Two, Three, Four...in order of position on the leader.

VC The communists of South Vietnam—a contraction of Viet Cong. They could be soldiers or civilians. The enemy; also gooks, the bad guys.

Vill Village. Also "Ville."

Wave-off Direction to discontinue a landing approach and go around for another try.

WIA Wounded in action.

Willy Pete White phosphorous warhead installed on a rocket or artillery round. Used by FACs to mark a target with a large ball of white smoke; easily identified by attack aircraft.

Willy the Whale Affectionate nickname for the F3D/EF10B.

Wing/MAW Marine Aircraft Wing, the basic element of Marine Aviation composed of several combat and support MAGs.

Wop The very distinguishing sound of an approaching Huey helicopter, as in "Wop, Wop, Wop, Wop, Wop"!

Made in the USA
Monee, IL
04 November 2026

67375608R00138